Wayne D. Monnery

CFD Simulation of Oilfield Separators

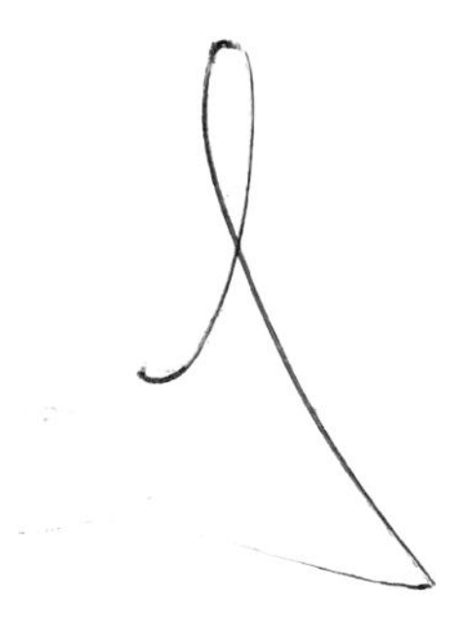

Ali Pourahmadi Laleh
William Y. Svrcek
Wayne D. Monnery

CFD Simulation of Oilfield Separators

A Realistic Approach

LAP LAMBERT Academic Publishing

Impressum/Imprint (nur für Deutschland/only for Germany)
Bibliografische Information der Deutschen Nationalbibliothek: Die Deutsche Nationalbibliothek verzeichnet diese Publikation in der Deutschen Nationalbibliografie; detaillierte bibliografische Daten sind im Internet über http://dnb.d-nb.de abrufbar.

Coverbild: www.ingimage.com

Verlag: LAP LAMBERT Academic Publishing GmbH & Co. KG
Heinrich-Böcking-Str. 6-8, 66121 Saarbrücken, Deutschland
Telefon +49 681 3720-310, Telefax +49 681 3720-3109
Email: info@lap-publishing.com

Approved by: Calgary, The University of Calgary, PhD Thesis, 2010

Herstellung in Deutschland:
Schaltungsdienst Lange o.H.G., Berlin
Books on Demand GmbH, Norderstedt
Reha GmbH, Saarbrücken
Amazon Distribution GmbH, Leipzig
ISBN: 978-3-8454-0351-9

Imprint (only for USA, GB)
Bibliographic information published by the Deutsche Nationalbibliothek: The Deutsche Nationalbibliothek lists this publication in the Deutsche Nationalbibliografie; detailed bibliographic data are available in the Internet at http://dnb.d-nb.de.

Cover image: www.ingimage.com

Publisher: LAP LAMBERT Academic Publishing GmbH & Co. KG
Heinrich-Böcking-Str. 6-8, 66121 Saarbrücken, Germany
Phone +49 681 3720-310, Fax +49 681 3720-3109
Email: info@lap-publishing.com

Printed in the U.S.A.
Printed in the U.K. by (see last page)
ISBN: 978-3-8454-0351-9

To my mother, for her Pure Love

&

To my father, for his Profound Logic

APL

About the Authors

Ali Pourahmadi Laleh is an aspiring chemical engineer with more than 11 years of experience as a Research Engineer, involved in rectifying industrial scale process inefficiencies and optimizing chemical plants. His earned academic degrees include a BSc (2000) from Sahand University of Technology, an MSc (2003) from Sharif University of Technology, and a PhD (2010) from the University of Calgary, all in the field of Chemical Engineering. Through organized academic courses and an industrial training program in the Polyethylene Unit of the Tabriz Petrochemical Complex, his background in chemical engineering has been formed with a strong orientation toward petrochemical and oil production processes. In the MSc program, with process design specialization, he proposed a novel approach for automatic design of the optimum distillation column sequence using a global optimization method, i.e. Genetic Algorithms. In the PhD career, he developed an efficient strategy for realistic simulation of oilfield separators. As will be explained in this book, the simulation was based on Computational Fluid Dynamics (CFD) and led to rectifying the design issues with these multiphase separators. He also performed an industrial research on coal liquefaction in summer 2008. Dr. Pourahmadi Laleh served as a Teaching Assistant at the University of Calgary and hold Ursula & Herbert Zandmer award from the university for the academic years of 2006-2009. He is currently working as a Research Engineer in the Reservoir Simulation Group directed by Dr. John Chen at the University of Calgary. The simulation case of interest is in-situ upgrading process by hot fluid injection for Athabasca oilfield. This research involves feasibility and sensitivity analyses of the mentioned process and will provide its optimized parameters. Dr. Pourahmadi Laleh is a member of the Canadian Society for Chemical Engineering.

William Y. Svrcek is a Professor Emeritus (retired, 2009) of Chemical and Petroleum Engineering at the University of Calgary, Alberta, Canada. He received his BSc (1962) and PhD (1967) degrees in Chemical Engineering from the University of Alberta, Edmonton. Prior to joining the University of Calgary he worked for Monsanto Company as a senior systems engineer and as an Associate Professor (1970-75) in the Department of Biochemical and Chemical Engineering at the University of Western Ontario, London, Ontario. Dr. Svrcek's teaching and research interests center on process simulation control and design. He has authored or coauthored over 200 technical articles/reports and has supervised over 50 graduate students. He has been involved for many years in teaching the continuing education course titled "Computer Aided Process Design - Oil and Gas Processing" that has been presented worldwide. This course was modified to include not only steady-state simulation but also dynamic simulation and control strategy development and verification. Dr. Svrcek was also a senior partner in Hyprotech, now part of Aspen Technology, from its incorporation in 1976. As a Principal, Director, and President (1981-1993) he was instrumental in establishing Hyprotech as a leading international process simulation software company. He is currently providing leadership and vision in process simulation software as the President of Virtual Materials Group Inc. He is a registered Professional Engineer in Alberta, and a member of professional societies that include The Canadian Society for Chemical Engineering, American Institute for Chemical Engineers and the Gas Processors Association of Canada.

Wayne D. Monnery, PhD, P.Eng., has 24 years of industrial experience as a Process Engineer, with recognized expertise in applied thermodynamics, process simulation and physical properties of petroleum systems, as well as in sweet gas processing, sour gas treating and sulfur recovery. He has also worked on heavy oil and steam assisted gravity drainage (SAGD) facility simulation and design. Dr. Monnery has a PhD in Chemical Engineering from the University of Calgary, Canada, 1996. His PhD work involved developing a viscosity prediction model using statistical thermodynamics. He is also an Adjunct Associate Professor in the Department of Chemical and Petroleum Engineering at the University of Calgary and has done research in sulfur plant kinetics, water content of high pressure acid gases for acid gas injection and phase separation. He has also done industrial research in alternative sour gas treating. He has taught courses in gas processing at the University of Calgary, for the Institute of Gas Technology, USA, for UIS/Ecopetrol in Colombia, SA. and for UNI, Lima, Peru. Dr. Monnery is President of Chem-Pet Process Tech Ltd., which specializes in oil and gas facility design, process trouble-shooting and debottlenecking.

Preface

Multiphase separators are generally the first and largest process equipment in an oil production platform and are one of the most prevalent unit operations in any process. Furthermore, this primary separation step is a key element in oil and gas production facilities in that all downstream equipment, such as compressors, are completely dependent on the efficient performance of these multiphase separators.

The literature on this critical unit operation abounds with macro studies and design methodologies for two and three-phase vertical and horizontal separators, though they often lack validation of their recommended parameters. There are very few studies/papers that provide the micro details of the actual separation process. In order to deal with this shortcoming, we undertook a research project whose purpose was set to use Computational Fluid Dynamics (CFD) as a method to simulate the detailed performance of multiphase separators. The research project involved the CFD simulation of four pilot-plant-scale two-phase separators and one industrial scale three-phase separator, including all the installed internals to verify the developed models versus actual separation data. Realizing the difficulties involved with CFD modeling of multiphase fluid flows and the required high-quality simulation of multiphase separator performance, a proven commercial CFD package, Fluent 6.3.26, was used. In order to capture both macroscopic and microscopic aspects of multiphase separation phenomenon, an efficient combination of two multiphase models was used. The Volume of Fluid (VOF) model was used to simulate the phase behavior and fluid flow patterns, and the Discrete Phase Model (DPM) was used to model the movement of fluid droplets injected at the separator inlet. The "particle tracking" based simulation of the multiphase separation process was the key aspect of this research project, and the developed model did provide high-quality visualization of multiphase separation process.

The objectives, aside from developing the CFD model, were to use the simulation results to validate and provide additional criteria for separator design. These criteria were combined with an industry standard algorithmic design method to specify a realistic optimum separator design. For this purpose, a useful method was developed for estimation of the droplet sizes used to calculate realistic separation velocities for various oilfield conditions. The velocity constraints caused by re-entrainment in horizontal separators were also studied via comprehensive CFD simulations, and led to novel correlations for the re-entrainment phenomenon.

This book has been organized as follows. Chapter One provides the background and the objectives of the research project. Since readers tend to skip over the introductory material and focus on the novel achievements of a research study, only the most useful comments, which are necessary to set the stage for the reminder of the book, have been presented in the first chapter. Chapter Two presents a detailed literature review of the very few CFD based studies performed on multiphase separators. Moreover, the classic guidelines for separator design are also reviewed, providing a useful literature review within the research project scope. Chapter Three provides the details of the developed procedure for the CFD models required in the Fluent software for simulating the complex features of multiphase separators. In contrast with the published CFD based studies, all details of the CFD modeling phase will be provided. Chapter Four presents the results of using the developed CFD models of Chapter Three for the simulated separators. In Chapter Five, the effective model assumptions and settings are used to establish new/improved separator design criteria. Finally, Chapter Six presents the conclusions and recommendations of this research project.

The supplementary package of the book such as the fluid flow profiles, and the separation efficiency plots can be downloaded from internet. For a download link, please email us at cfdsoos@gmail.com. Furthermore, we are looking forward to hearing from you with any feedback on the book or any comment for improving its contents: your ideas are very important to us.

Ali Pourahmadi Laleh

William Y. Svrcek

Wayne D. Monnery

Acknowledgements

This book has been written based on a significant research project which was developed at the University of Calgary. I do believe that this project could not be so successful without instrumental support of so many people. Here, I would like to highlight the most important ones and acknowledge their contribution. In this regard, my sincere thanks go to Dr. William Y. Svrcek for his persuasive and powerful character which made it possible to accomplish this research project. Indeed, I am very proud of being his last PhD student. I have learned a lot from his efficient and superb approaches to directing my research. His precious moral and financial supports are gratefully appreciated.

I also express my gratitude to Dr. Wayne D. Monnery who contributed greatly to this project. His excellent insights and recommendations were extremely conducive in initiating and shaping this research.

I am in debt of Dr. Ramin B. Boozarjomehry, my supervisor in the MSc career, for his everlasting encouragement and technical supports during my PhD career. His significant role in my graduate career is most appreciated, and I wish all the best for him for good.

During the initial periods of my research, I took advantage of very helpful discussions with Dr. Jalel Azaiez. I do appreciate him for providing me with good ideas and also useful documents on the multiphase flow issues.

I would also thank Dr. Apostolos Kantzas as I have used his software license while developing my CFD models. In addition, technical support of his research assistant, Blake Chandrasekaran, is appreciated.

I am grateful to Dr. Thomas G. Harding for making the "Ursula & Herbert Zandmer Graduate Recruitment" scholarship possible to me in consecutive years of 2006 to 2009. This award played a major role in reaching my academic objectives.

As an expert on the computational resources, Dr. Doug S. Phillips helped me several times to overcome the difficulties arisen while submitting/performing parallel simulation cases on the WestGrid environment. His great responsibility and adorable diligence are gratefully acknowledged.

I am indebted to my parents and sisters who set the stage for me to pursue my studies in Canada. I extend my grateful appreciations to them for their encouragement, patience, and support.

My wife, Roza Kazemi, joined the scenario in early 2008, and since then, she has been taking all the proactive measures to my success in academic and personal life. Her invaluable contribution is sincerely acknowledged.

Ali Abbaspour, Hassan Abbasnezhad, Ehsan Aminfar, and Ali Yazdani, with their lovely personality, have been more than friends to me since I had a chance to have them in my life. I would like to appreciate these gentlemen for the moral assistance they have provided always.

Finally, I like to appreciate the administration staff of the Chemical & Petroleum Engineering department of the University of Calgary for their technical support. Particularly, I am very thankful to Andrea Cortes for her great personality and extraordinary responsibility.

Ali Pourahmadi Laleh

Contents

List of Tables

List of Figures and Illustrations

List of Abbreviations and Nomenclature

Abbreviations

API	American Petroleum Institute
CFD	Computational Fluid Dynamics
DPM	Discrete Phase Model
FLOSS	Flow Simulator for Separators
FPSO	Floating Production Storage and Offloading
FWKO	Free Water Knock Out
GPSA	Gas Processors Suppliers' Association
LDA	Laser Doppler Anemometry
NATCO	National Tank Company
PC	Personal Computer
PDA	Phase Doppler Analysis
PISO	Pressure-Implicit with Splitting of Operators
PR	Peng Robinson
RAM	Random Access Memory
SIMPLE	Semi-Implicit Method for Pressure-Linked Equations
SINTEF	"Stiftelsen for industriell og teknisk forskning" which means: Foundation for Scientific and Industrial Research
VDM	Visual Dynamic Modeling
VOF	Volume of Fluid

Nomenclature

a	distance between holes of a perforated plate (Figure 3-11), m
A	surface area required for liquid-liquid separation or vessel cross-sectional area, m^2
A_f	open area of a perforated plate, m^2
A_H	vessel cross-sectional area for heavy liquid phase, m^2
A_{HLL}	surface area corresponding to high liquid level, m^2
A_L	vessel cross-sectional area for light liquid phase, m^2
A_{LLL}	surface area corresponding to low liquid level, m^2
A_{LLV}	surface area of light liquid above vessel bottom, m^2
A_p	total area of a perforated plate, m^2
A_V	surface area of vapor disengagement zone, m^2
b_i	body force per unit mass in i direction, N/kg
C	discharge coefficient for a perforated plate
C_2	inertial resistance factor, m^{-1}
C_D	drag coefficient
C_{ij}	inertial resistance factor in j direction exerted on a plane perpendicular to i axis, m^{-1}
d	diameter of holes in a perforated plate, m
$\overline{d}$	volume mean diameter in Rosin-Rammler equation, μm, m
d_{eff}	efficient diameter of droplet distribution, μm

d_{min}	minimum droplet diameter, μm
d_{max}	maximum droplet diameter, μm, m
d_p	particle diameter, μm, m
D	diameter of separator or internal diameter of a pipe, m
D_B	boot diameter, m
D_{ij}	viscous resistance factor in j direction exerted on a plane perpendicular to i axis, m^{-1}
f	friction factor
g	gravity acceleration, m/s^2
H	liquid phase thickness, m
H_H	thickness of heavy liquid phase, m
H_{HLL}	height of high liquid level, m
H_L	thickness of light liquid phase, m
H_{LLB}	light liquid height in boot, m
H_{LLL}	height of low liquid level, m
H_{LLV}	light liquid height in separator, m
$HolePitch$	diagonal distance between holes of a perforated plate, m
H_R	height of light liquid phase above the outlet, m
H_S	surge height, m
H_{sep}	separation height in horizontal separators, m
H_T	total height of a vertical separator, m
H_V	height of vapor disengagement zone, m
H_W	weir height, m
I	turbulence intensity
k	thermal conductivity, $W/m°C$
K	settling velocity coefficient, m/s
L	length of separator or residence length, m
L_{min}	minimum length required for vapor-liquid separation, m
$\dot{m}$	mass flow rate through a perforated plate, kg/s
m_{in}	mass of droplets entered to the demister, kg
m_{out}	mass of droplets exited from the demister, kg
n	spread parameter in Rosin-Rammler equation
N_μ	interfacial viscosity number as per Equation 5-18
P	operating pressure, kPa
$\dot{q}$	rate of volumetric heat addition per unit mass, W/kg
Q	volumetric flow-rate of liquid phase, m^3/s
Q_c	continuous phase flow-rate, m^3/s
Q_d	dispersed phase flow-rate, m^3/s
Q_{HL}	heavy liquid phase flow-rate, m^3/s
Q_{LL}	light liquid phase flow-rate, m^3/s
Q_V	vapor phase flow-rate, m^3/s

Re	Reynolds number
Re_p	particle Reynolds number
S_i	source term for the momentum equation in i direction, *Pa/m*
t_{HL}	separation time for heavy liquid droplets from light liquid phase, *s*
$t_{interface}$	time required for droplet penetration through the interface, *s*
t_{LH}	separation time for light liquid droplets from heavy liquid phase, *s*
T	temperature, *K*
T_r	reduced temperature
u	velocity component in x direction, *m/s*
U	internal energy, *J*
U_d	separation velocity of dispersed phase, *m/s*
v	velocity component in y direction, *m/s*
V	velocity, *m/s*
V_B	velocity of heavy liquid phase in the boot, *m/s*
V_c	superficial velocity of continuous phase, *m/s*
V_H	holdup volume, m^3
V_S	surge volume, m^3
V_{sep}	relative efficient separation velocity in a separator, *m/s*
$V_{sep,vertical}$	apparent efficient separation velocity in a vertical separator, *m/s*
V_V	velocity of vapor phase, *m/s*
$V_{V,\max}$	maximum safe velocity of vapor phase to avoid re-entrainment, *m/s*
$V_{Water,\max}$	maximum safe velocity of water phase to avoid re-entrainment, *m/s*
w	velocity component in z direction, *m/s*
We	Weber number
We_{crit}	critical Weber number as per Equation 3-25
We'_{crit}	critical Weber number in Levich theory as per Equation 3-30
Y_d	is the mass fraction of droplets with diameter greater than d

Greek Letters

α	permeability factor, m^2
β	volume fraction of dispersed phase
δ	thickness of a perforated plate, *m*
ΔH	height difference between the light and heavy liquid weirs, *m*
ΔP	pressure gradient, *Pa*
Δx	gradient along thickness of a perforated plate, *m*
ε	porosity of a perforated plate
φ	function of d_p as per Equation 4-3
ϕ	liquid dropout time from vapor phase, *s*
μ_c	continuous phase viscosity, *Pa.s*
μ_m	mixture viscosity as per Equation 4-2, *Pa.s*

μ_{Oil}	oil phase viscosity, *Pa.s*
μ_V	vapor phase viscosity, *Pa.s*
μ_W	water phase viscosity, *Pa.s*
ν	kinematic viscosity, m^2/s
ρ	density, kg/m^3
ρ_c	continuous phase density, kg/m^3
ρ_d	dispersed phase density, kg/m^3
ρ_H	heavy liquid phase density, kg/m^3
ρ_L	(light) liquid phase density, kg/m^3
ρ_{Oil}	oil phase density, kg/m^3
ρ_V	vapor phase density, kg/m^3
ρ_W	water phase density, kg/m^3
ψ	energy dissipation rate per unit mass, *W/kg*
σ	surface tension, *N/m*
τ	dynamic pressure fluctuation, *Pa*
τ_{ij}	stress in j direction exerted on a plane perpendicular to i axis, *Pa*
θ_H	residence time of the heavy liquid phase, *s*
θ_L	residence time of the light liquid phase, *s*

Subscripts

B	boot
c	continuous
crit	critical
d	dispersed, a diameter
D	drag
eff	efficient
H	heavy liquid, holdup
HLL	high liquid level
in	entered to
L	liquid, light liquid
LLL	low liquid level
m	mixture
min	minimum
max	maximum
out	exited from
p	particle
r	reduced
S	surge
sep	separation
T	total
V	vapor
W	water

Chapter One: Introduction

1.1 Gas-Liquid Phase Separators

Once a crude oil has reached the surface, it must be processed so it can be sent either to storage or to a refinery for further processing. In fact, the main purpose of the surface facilities is to separate the produced multiphase stream into its vapor and liquid fractions. On production platforms, a multiphase separator is usually the first equipment through which the well fluid flows followed by other equipment such as heaters, exchangers, distillation columns etc. Consequently, a properly sized primary multiphase separator can increase the capacity of the entire facility.

The oil as produced at the platform varies significantly from oilfield to oilfield. In some oilfields, water (brine) is not produced together with oil, and hence only the gas and the oil need to be separated (two-phase separation). However, usually, three-phase separation of oil, water, and gas is required in order to prepare the produced multiphase fluid for downstream processing.

1.1.1 *Multistage Separation*

A low-pressure crude oil with a low gas to oil ratio can be passed directly into a "flow tank" operating at atmospheric pressure in which the separation of the oil from other phases can take place in one stage (Skelton, 1977), that is "single-stage separation". Single-stage separation was at one time also applied to high pressure crude oils with high gas to oil ratio, but this process had several disadvantages such as a requirement for a very large separation vessel, the loss of a valuable fraction of oil in the gas phase, and extra mist and foam caused by abrupt pressure drop in the multiphase separator (Skelton, 1977). These drawbacks can be overcome by separating the phases from each other in a number of succeeding stages, "multistage separation". In this process, a number of separators are used in series, and the multiphase flow passes from one separator to the next while undergoing a controlled reduction of pressure. The intermediate operating pressures are specified so as to obtain a maximum recovery of liquid hydrocarbons and, at the same time, to provide maximum stabilization of the liquid and gas products. The optimum operating pressure of a separator is either calculated theoretically by flash calculations and engineering judgment or determined by field tests so that the pressure profile results in the highest economic yield from the sale of the liquid and gas hydrocarbons (Smith, 1987; Arnold and Stewart, 2008). Although, in the past, as many as seven stages of separation were used, the number has decreased to a maximum of five in current surface production operations (Skelton, 1977). The optimum number of separation stages can be determined by field testing or equilibrium calculations based on laboratory tests of the well fluid, and the economical number of stages are typically three or four, but five or six will pay out under special conditions (Smith, 1987).

1.1.2 *Multiphase Separator Terminology*

Multiphase separation can be carried out through various oil processing equipment with the specific terminology corresponding to each system. Hence, it is worth defining the most important multiphase separators in the oil industry before proceeding. The conventional oil and gas separator, which is normally installed on a production facility or platform, may be referred to as "oil and gas separator", "separator", "stage separator", or "trap". A "knockout vessel" is used to remove either water or all liquid from the well fluid flow. An "expansion vessel" is the first stage separator vessel usually operated at a low temperature. A "flash chamber" or "flash vessel" normally refers to a conventional oil and gas separator operated at low pressure as the second or third stage of the multistage separation. A "gas scrubber" is an oil and gas separator with a high gas to liquid ratio. In a "wet-type gas scrubber", dust, rust, and other impurities of the gas phase are washed using a bath of oil or other liquid, and the gas flows through a demister to further remove liquid droplets from the gas stream. A "dry-type gas scrubber" or "gas filter" is equipped with demisters and other coalescing media to aid in the removal of most of the liquid from a gas stream.

1.1.3 *Multiphase Separators*

The original phase separator was inclined, very long, and without any internal separating aids. During the retention time (around one minute), the gas and oil underwent a very limited separation. Sir Stephen Gibson designed this simple phase separator and also the multi-stage separation process. He was the first to put this process into operation in 1930 at the Haft-Kel oilfield in Iran (Skelton, 1977).

From this very simple separator, a wide variety of separator designs and configurations for multiphase separators in both vertical and horizontal orientations have been developed. Various parameters such as space and operating restrictions, oilfield variations, potential contaminants, and economic evaluations are considered in the design of a multiphase separation system. For instance, some separators may be equipped with special impingement internals to aid the separation process. Figure 1.1 shows some different common designs composed of "simple", "boot", "weir", and "bucket and weir". These designs offer a variety of methods to control the interface level in horizontal three-phase separators. "Simple" design separator can easily be adjusted to handle unexpected changes in oil or water density or flow rates (Arnold and Stewart, 2008). A boot, typically, is used when the water fraction is not substantial (less than 15-20% of total liquid by weight), and a weir is used if the water fraction is substantial (Monnery and Svrcek, 1994). The bucket and weir design is usually used when interface level control is difficult either in heavy oil applications or because of emulsions or paraffin problems (Arnold and Stewart, 2008).

In spite of the variety of design configurations proposed for multiphase separators, the phase separation process is accomplished in three zones: The first zone, primary separation, uses an inlet diverter so that an abrupt change in flow direction and velocity causes the largest liquid droplets to impinge on the diverter and then drop by gravity. In this zone, the bulk of the liquid phase is separated from the gas phase. In the next zone, secondary separation zone, gravity separation of fine droplets occurs as the vapor and liquid phases flow through the main section of the separator at relatively low velocities and little turbulence, and the liquid droplets settle out of the gas stream due to gravity. The liquid collection section in the bottom half of separator provides the retention time required for entrained gas bubbles or other liquid droplets to join their corresponding phases because of gravity and buoyancy. This section also provides the holdup and surge volumes for safe and smooth operation of the separator. Gas flows above the liquid phase while entrained small liquid droplets are again separated by gravity. The final zone, coalescing media, is designed for mist elimination in which very fine droplets that could not be separated in the gravity settling zone are separated by passing the gas stream through a mist eliminator. In this zone, vanes, wire mesh pad, or coalescing plates may be used to provide an impingement surface for very fine droplets to coalesce and form larger droplets which can be separated out of gas stream by gravity.

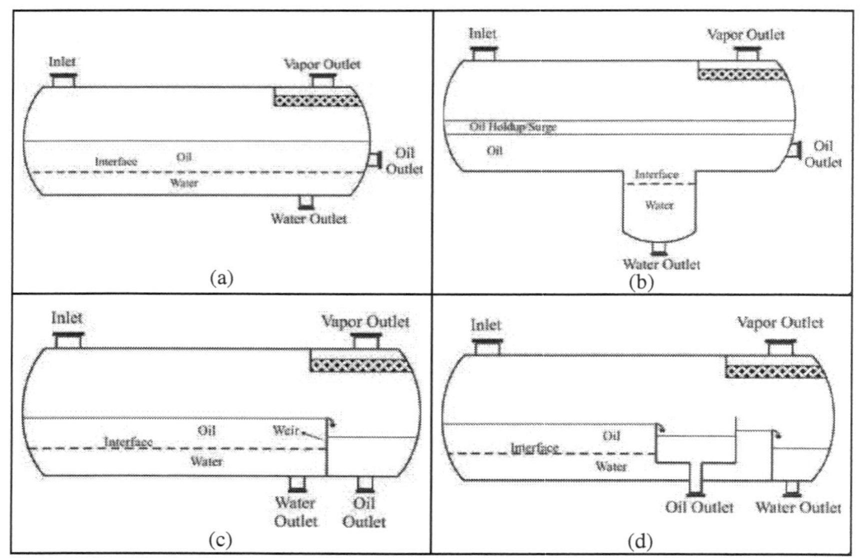

Figure 1.1. Different Common Designs of Horizontal Three-Phase Separators; (a) "Simple", (b) "Boot", (c) "Weir", and (d) "Bucket and Weir".

1.1.4 *Operating Pressures and Capacities*

The operating pressure of separators may vary from a high vacuum to around 35 *MPa* (Smith, 1987) and their capacities may range from a few hundred barrels per day to 100000 barrels per day or more (Skelton, 1977). As the operating pressure of a separator increases, the density difference between the liquid and gas phases decreases. Therefore, it is desirable to operate multiphase separators at as low a pressure as is consistent with other process conditions and requirements. Most multiphase separators operate in a pressure range of 138 *kPa* to 10340 *kPa* (Smith, 1987).

1.1.5 *Separator Internals*

The modern separator designs have much higher capacities than the original type and are very short in length due to internals such as inlet diverters, controls, flow-distributing baffles, and mist extractors which enhance the separator efficiency. A "weir" design horizontal three-phase separator and a vertical three-phase separator, both with the installed internals, are shown in Figure 1.2 and Figure 1.3, respectively. As represented, both separators are equipped with inlet diverters, controls, wire mesh demisters and pressure relief devices. In the vertical separator, a chimney is used to equalize gas pressure between the lower and upper sections of the vessel. Important common internals are explained in the following subsections.

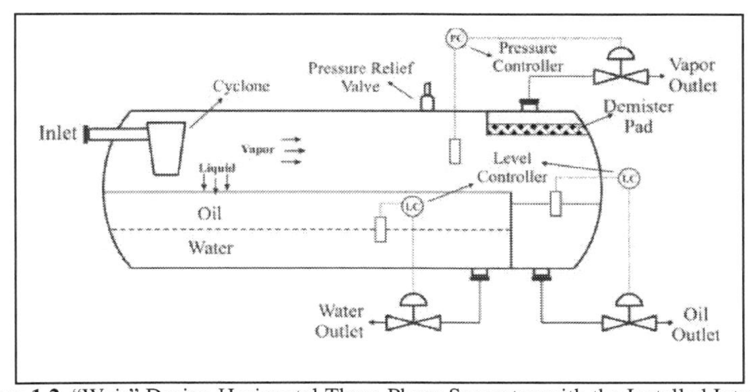

Figure 1.2. "Weir" Design Horizontal Three-Phase Separator with the Installed Internals.

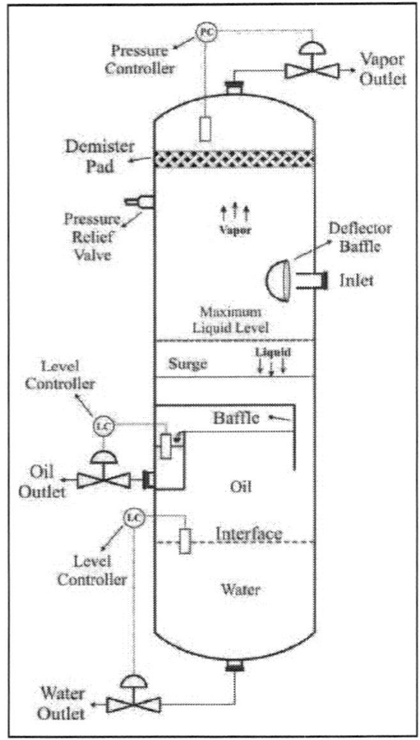

Figure 1.3. Vertical Three-Phase Separator with the Installed Internals.

1.1.5.1 Inlet Diverters

Usually a deflector baffle or a cyclone is used as inlet diverter in the separator. Deflector baffles come in various shapes and can be installed at different angles. However, hemisphere or conical designs are preferred because they cause fewer disturbances than plates or angle iron and reduce re-entrainment and emulsion problems (Arnold and Stewart, 2008). Cyclone diverters are increasingly used in oil production facilities as they promote foam breaking and mist elimination while performing the bulk gas-liquid separation in the inlet zone (Chin et al., 2002). Hence, cyclone diverters can be used to increase the operating capacity of multiphase separators.

1.1.5.2 Controls

Separators are required to operate at a predetermined pressure which is specified by economic and engineering studies. The fixed operating pressure in a separator is achieved by using an automatic back pressure regulator on the gas outlet line. This device maintains a steady operating pressure in the vessel. The liquid level controllers and liquid outlet control valves are used to maintain constant oil and water levels in the separator. Consequently, the operation of modern separator systems is completely automatic.

1.1.5.3 Mist Eliminators

Droplets with diameters of 100 μm and larger will generally settle out of the gas stream in most average-sized separators. However, mist eliminators are usually required to remove smaller droplets from the gas phase (Smith, 1987). As 95% of droplets entrained in the gas stream can be separated in economically-sized separators without coalescing media, the efficiency can be increased to around 100% by installing mist eliminators (Walas, 1990; Sinnott, 1997; Arnold and Stewart, 2008). These coalescing media can consist of a series of vanes, a knitted wire mesh pad, or cyclonic passages to remove the very fine droplets from the gas phase by impingement on a large surface area where they collect and collide with adjacent droplets. The mechanisms used in various mist eliminators are gravity separation, impingement, change in flow direction and velocity, centrifugal force, coalescence, and filtering (Smith, 1987). Mist eliminators can be of many different designs exploiting one or more of these mechanisms.

1.1.5.3.1 Wire Mesh Demisters

The wire mesh pads, shown in Figure 1.4, are made of knitted wire mesh and are installed by a lightweight support inside separators. Wire mesh pads have increasingly been used for mist elimination in separators. Since the early 1950s, the wire mesh demisters have been used in natural gas processing with the main use of removing fine droplets ranging from 10 to 100 μm in diameter from the gas stream (Smith, 1987). Wire mesh demisters are still frequently used in the natural gas industry and, with proper design, can separate even very fine droplets which are less than 10 μm in diameter from the gas stream (Fewel and Kean, 1992; Sinnott, 1997). Generally, high separation efficiencies at a low capital and maintenance cost are experienced from using standard wire mesh demisters. The pressure drop is a function of the entrainment load, the mesh pad design, and gas velocity but does not usually exceed 295 *Pa* (Lyons and Plisga, 2005). Mesh pad manufacturers report excellent performance from 30% to 110% (with 75% as the preferred operating value) of their recommended velocity (Evans, 1974; Walas, 1990).

Any common metal, such as carbon steel, stainless steel and aluminum, or plastic material may be used in these devices. During 1960-1970, considerable development effort was made to improve the performance of wire mesh demisters by using a combination of filaments of different materials and diameters. These types of wire mesh demisters do provide improved separation capacities at a cost of lower operating range as they flood at velocities 50% below those of regular demisters (Smith, 1987).

Although pad thicknesses up to 0.9 *m* have been used, a pad thickness of 0.10 *m* to 0.15 *m* is normally sufficient for most separator applications (Gerunda, 1981; Walas, 1990; Lyons and Plisga, 2005). However, the "fouling" tendency of wire mesh demisters may restrict their applications to gas scrubbers (Smith, 1987). In fact, knitted wire mesh may foul or plug from paraffin deposition and other impurities and thus reduce separation efficiency dramatically after a short period of service. In such cases, vane-type or centrifugal demisters are used.

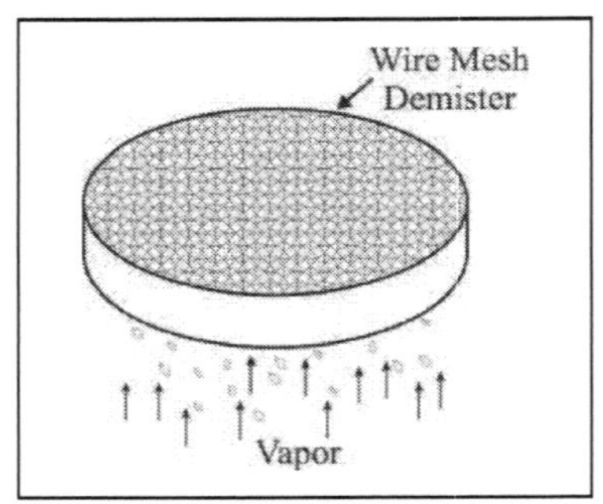

Figure 1.4. Wire Mesh Pad Mist Eliminator.

1.1.5.3.2 Vane-Type Demisters

Vane-type demisters, Figure 1.5, are widely used in oil and gas separators. The separation mechanisms used in most of vane-type demisters are impingement, change in flow direction and velocity, and coalescence. Vane-type demisters use the inertia of the liquid droplets in the gas stream to collect a film of liquid on the vane surface. Vane-type demisters are inexpensive and usually will not plug or foul with paraffin or other contaminants, hence, providing a good separation performance under widely changing field conditions (Smith, 1987). Pressure drop across the vane-type demisters are very low, ranging from 250 *Pa* to 1 *kPa* (Smith, 1987). In order to establish a required minimum pressure drop, vane-type demisters are sized by their manufacturers (Arnold and Stewart, 2008). It is usually preferred to use vane-type demisters for most oil and gas separators, particularly in high-capacity applications, because they can significantly reduce the separator size (Fewel and Kean, 1992).

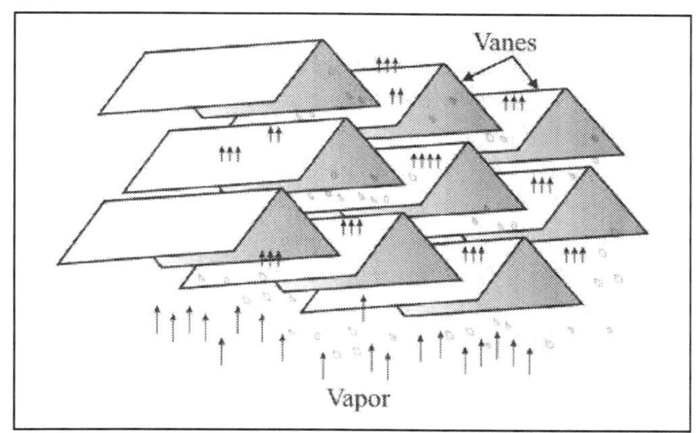

Figure 1.5. Vane-Type Mist Eliminator.

1.1.6 *Operational Difficulties*

The most common factors which can reduce separator performance are very high or very low liquid level, level control failure, improper design, damaged vessel internals, foam, vortex formation in liquid outlet zones, plugged liquid outlets, and exceeding the design capacity of the vessel (Arnold and Stewart, 2008).

The common approaches used for improving the separator performance in difficult cases, as proposed by Blezard et al. (2000), are increasing droplet size of dispersed phase (e.g., by promoting coalescence), inducing a high acceleration on droplets (e.g., by using centrifugal force), increasing the difference between fluid densities (e.g., by introducing diluents), and decreasing the viscosity of the liquid phases (e.g., by heating). Some of these approaches may be combined to overcome a difficult separation task. In the following, some different measures taken for operating foamy, emulsified, or contaminated crude oils are outlined.

Foamy crude oils hinder liquid level control and also reduce the separation space of the separator. To improve the separator performance, it is usually advantageous to inject a silicon defoaming agent into the foamy oil stream (around 1×10^{-6} m^3 for 1 m^3 of oil) before it enters the separator (Skelton, 1977). This agent breaks up the foam and keeps oil from being carried over by the gas phase, leading to an effective increase in the capacity of the separator. The other approaches that assist in breaking the foam are settling, baffling, heat, and centrifugal force (Smith, 1987). For separators suffering from liquid carryover while processing foamy crude oils, or glycols, amines, and similar materials (with high foaming tendency), a dual mist eliminator system composed of a vane-type demister at a lower level and a wire mesh pad at higher level with a gap of 0.15 to 0.30 m between them is usually used (Lyons and Plisga, 2005).

The other separation difficulty is caused by thoroughly emulsified oil. The water phase enters the bottom of the producing well and usually breaks up into fine droplets on its way to the surface. These fine droplets form an emulsion with the oil phase which can lead to a fully emulsified phase. Separation of the thoroughly emulsified phase is extremely difficult, and it is often recommended that as much of the water as possible be removed at the well head (Skelton, 1977). This treatment is done by processing the crude oil through a large vessel (long retention time) which would allow the larger water droplets to settle out. If further treatment of the separated oil phase is necessary, the oil phase may be heated to help break down the emulsion. Usually, a surface tension reducing chemical is also added to enhance the treatment. Generally the combined application of heat and chemicals is sufficient to reduce the water and, consequently, the salt content of the oil phase to an acceptable level (Skelton, 1977). However, sometimes the use of an electrical coalescing media may be necessary to achieve specification level of oil product. In this treatment method, the oil-water stream is exposed to an electrical field which agitates the water droplets and causes them to collide and coalesce into larger droplets and then to settle out of the oil phase. The resultant water concentration in the effluent oil stream is usually less than 0.5% (Blezard et al., 2000).

Another emulsion problem is experienced when some very fine particles cause a stabilized rag layer at the oil-water interface. In order for the separator to operate properly, the rag layer must be regularly broken or removed. For this purpose, some techniques such as filtration, heating, and chemical injection are used (Hooper, 1997).

Contaminated crude oils are also difficult to process. The most common contaminants are sand, silt, mud, and salt. Medium-sized sands in small quantities can be removed by an oversized vertical settler. The residue should be removed periodically by draining from the vessel bottom. Salt may be removed by washing the oil with water and then separating the salty water from the oil phase.

1.2 CFD as a Separator Modeling Method

Computational Fluid Dynamics (CFD) is inherently connected with the "fluid" concept. It is interesting that this "fluid" concept can still be defined as Isaac Newton proposed more than 300 years ago in the following elegant way: "A fluid is any body whose parts yield to any force impressed on it, and by yielding, are easily moved among themselves." The physical features of any fluid flow are governed by three fundamental physical principles: mass is conserved, Newton's second law applies, and energy is conserved. These fundamental physical principles can be represented in terms of mathematical equations, generally in the form of integral equations or partial differential equations. Computational fluid dynamics is the art of replacing the integrals or the partial derivatives in these equations with their equivalent discretized algebraic forms. These discretized algebraic equations are then solved to provide numbers for the flow field values at discrete points in time and/or space. Therefore, in contrast with an analytical solution, the final product of a CFD modeling is a collection of numbers. CFD solutions generally require the iterative manipulation of many thousands, even millions, of numbers. This task is obviously impossible without the aid of a high-speed digital computer which accelerated the practical development of CFD. The historical development of CFD, as reviewed by Anderson (1995), indicates that before 1970, there was no CFD in the way that we think of it today, and although there was CFD in 1970, the storage and speed capacity of computers limited all practical solutions essentially to two-dimensional flow problems. However, by 1990, this story had changed dramatically. In today's CFD modeling applications, three-dimensional flow field solutions are abundant and such solutions are becoming more and more prevalent within industry and government facilities. Indeed, some computer programs for the calculation of three-dimensional flows have become industry standards, resulting in their use as a tool in the design process. In short, CFD, along with its role as a research tool, is playing an increasingly stronger role as a design tool.

The high storage capacities and calculation speed of present computers and advanced techniques devised in modern CFD solvers have culminated in today's common use of commercial software packages for CFD simulation of industrial equipment. Currently, CFD software packages are routinely used to modify the design and to improve the operation of most types of chemical process equipment, combustion systems, flow measurement and control systems, material handling equipment, and pollution control systems (Shelley, 2007). For implementation of a CFD simulation using a commercial software package, the geometry of the object of interest is specified (with a computer aided design drawing of the object) and the corresponding discretized grid system is created using a mesh-generation tool. For mesh generation, present software tools provide some predefined building units in a variety of forms such as tetrahedral, pyramidal, hexahedral, and recently, polyhedral blocks. However, generating a high quality mesh for the system is still one of the most technical and time-consuming phases in any CFD based analysis. After preparing the grid system, the initial and boundary conditions of the problem are specified, and the CFD parameters are set. Finally, the CFD software proceeds with the iterative process of solving the fundamental equations for fluid flow. As noted, once a converged solution is achieved, CFD simulation output is a collection of numbers which correspond to the defined points in space or time. In order to visualize these CFD simulation results and obtain qualitative aspects of the system, the post-processing tool of CFD software is used.

CFD complements the approaches of pure theory and pure experiment in the analysis and solution of fluid dynamic problems. Although CFD will probably never completely replace either of these approaches, it helps to interpret and understand the results of theory and experiment. It should be noted that suitable problems for CFD often involve predictions outside the scope of published data, where experimental studies are too expensive or difficult or where the development of an insight is required (Sharratt, 1990). Therefore, CFD is primarily an insight tool which is useful for understanding the important features of a system and for elucidating and solving some system uncertainties and problems.

1.3 Motivations for the Performed Research Project

Many industries have a variety of processes in which multiphase mixtures must be separated. This is the situation in the hydrocarbon production, in which produced oil, water and gas must be separated. Optimum and efficient design of the multiphase separators, which are generally the first and largest elements of process equipment in an oil production platform, will reduce capital cost and will improve operating performance of the hydrocarbon production facility. In fact, the multiphase separation process is a key element in the oil and gas production facilities in that downstream equipment, such as compressors, are completely dependent on the efficient and proper performance of the multiphase separators. Thus, multiphase separators play a critical role in onshore and offshore oil production. Figure 1.6 shows a schematic of a typical three-phase separator, in which a mixture of oil, gas and water enters the separator as a high momentum jet and proceeds through a cyclone designed to perform a bulk gas-liquid separation. The liquid then falls into a pool in the lower part of the vessel and the gas together with the entrained hydrocarbon liquid drops flow to the upper section of the separator. The released gas rises to the outlet at the top of the vessel while oil and water flow to separate outlets at the bottom of the vessel. In the upper part of the vessel, oil and water droplets flow down to the liquid interface as the multiphase fluid flows to the outlet. Usually, there is also a demister pad which removes the very fine droplets of liquid from the gas by impingement on a surface where they coalesce. Several operational problems are often experienced that include emulsion problems inside the separator, water level control failure, a rising water content in oil exiting the separator because of increased production of water, oversized vessels, low separation efficiency, and so on.

This research project applies CFD based simulation to model two and three-phase separators. The CFD model is used to analyze velocity, pressure and concentration of multiphase fluid flow within the phase separation process in the separators. CFD, which involves the solution of the governing equations for fluid flow at thousands of discrete points on a computational grid in the flow domain, provides a unique picture of the fluid movement within the vessel. The provided fluid flow visualization helps to gain a better understanding of multiphase separation process and the corresponding design issues.

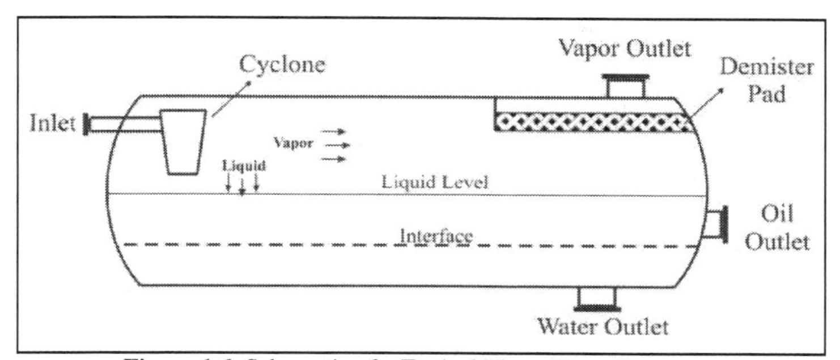

Figure 1.6. Schematic of a Typical Three-Phase Separator.

Although CFD simulation is not a substitute for experience, it does provide more information than can be obtained from physical experiments and can bring far more useful information to bear on the design and operation of the multiphase separation process. CFD provides a flexible and economical means of testing the performance of alternative designs or theoretical advances for cases that are very difficult or expensive to physically test by experimental studies. For example, the geometry of a vessel and its internals can be changed more easily and re-analyzed to determine the effect of the change. In addition, scaling is not an issue with CFD because the simulation can easily be adjusted to represent any equipment size. Therefore, CFD simulation is a useful tool that once validated can reproduce conditions that would be impossible or impractical to duplicate in physical testing. CFD studies can efficiently guide engineers to the sources of performance inefficiencies. By developing a detailed CFD computer model, a possible explanation for some of the problems is determined and possible solutions can then be proposed. This research illustrates the benefits that CFD analysis can provide in optimizing the design of new separators and solving problems with existing designs. Thus, there is a direct tangible benefit to industry.

Chapter Two: Literature Review

Multiphase separators and their performance are key issues to economical and stable hydrocarbon fluid processing. Therefore, the public literature and corporate literature are extensive, particularly regarding two-phase separators. Most of these documents propose separator design guidelines and some address operating performance issues associated with multiphase separators. In a very few studies, CFD based simulations have been performed to provide a realistic picture of fluid phase separation phenomena and to improve separator efficiency. These CFD based studies, however, have generally been focused on performance issues in specific operating separators.

The present chapter will present a review of the CFD based studies performed on multiphase separators. Furthermore, since this research project developed improved criteria for designing multiphase separators, classic guidelines for design of separators are also reviewed in this chapter to provide a comprehensive literature review within the research project scope.

2.1 Classic Methods for Design of Multiphase Separators

In classic methods, multiphase separators are designed so as to provide sufficient disengagement space for gas phase from which liquid droplets settle and to provide adequate retention time for liquid phase(s) to establish satisfactory gas-in-liquid or liquid-liquid separation (in three-phase operation). Some heuristics, rules-of-thumb, have also been suggested which are usually conservative and lead to oversized separators (Abernathy, 1993). In more systematic procedures, droplet settling theory is applied for evaluating vapor-liquid or liquid-liquid separation requirements, and adequate retention times may be assumed based on experience, scale model predictions, or field data. Using systematic procedures, a software package was developed for design of two-phase and three-phase separators (Grødal and Realff, 1999). The design procedure was based on droplet settling theory, and the optimum solution was determined by applying Sequential Quadratic Programming (SQP) techniques. As stated by Grødal and Realff (1999), the most comprehensive approach among the classic methods was proposed by Svrcek and Monnery in two technical papers (Svrcek and Monnery, 1993; Monnery and Svrcek, 1994). In this approach, based on accepted industrial guidelines, an algorithmic method was developed for designing the optimum (the most economical) multiphase separators through iteration (for horizontal separators) or height adjustment (for vertical separators). In order to provide more details of the approach, the main steps used in designing three-phase separators proposed by Monnery and Svrcek (1994) are presented in Appendix A.

2.1.1 *Common Design Aspects*

2.1.1.1 Vessel Orientation

The first design issue connected with multiphase separators is their orientation. Multiphase separators can be designed and installed in either horizontal or vertical orientations. Vertical separators occupy little plot space, and they can be more easily transported and installed than horizontal separators. However, other factors, such as high operating capacity, and processing capabilities, do require horizontal separators. Therefore, a choice is required for each design/processing situation. For instance, in the Middle East, the separators are usually horizontal, but in the USA, where the quantities of crude oil processed at each battery are much less than in the Middle East, vertical separators are more common (Skelton, 1977).

To aid the separator design process, rules of thumb have been developed. As concisely proposed by Evans (1974), vertical separators are used when there is small liquid load, limited plot space, or where ease of level control is desired.

The GPSA Engineering Data Book (1998) states while horizontal separators are most efficient for high capacity operations and where large amounts of solution gas are in the liquid phase, vertical separators are usually used if the gas to liquid ratio is high or total gas volumes are low.

Arnold and Stewart (2008) have shown that horizontal separators are more economical than vertical types and provide better processing/operation when emulsions, foam, or high gas to oil ratio fluids are present, but they have some limitations in sand and surge processing and require more plot space than equivalent vertical separators. Vertical separators are often preferred for operating either low or very high gas to oil ratio fluids.

A more comprehensive list of similar heuristics was proposed by Smith (1987). However, although the presented guidelines are useful, as emphasized by Svrcek and Monnery (1993), it is necessary for practical design of separators to compare both horizontal and vertical orientation designs to determine which is more economical.

2.1.1.2 The Aspect Ratio of a Separator

Although there is not a unique set of diameter and lengths of a separator that satisfies a given capacity requirement, the separator aspect ratio, defined as the ratio of length (or height) to diameter in a separator (L/D), should be in a reasonable range which is specified by economic analyses. Plot restrictions may also dictate the separator aspect ratio. Economic studies have led to the following heuristics for separator aspect ratio.

Smith (1987) suggested applying a minimum of about 1.0 to 2.0 and a maximum of about 8.0 to 9.0 to the aspect ratio. The preferred range of the aspect ratio for horizontal separators is 2.0-6.0, and the value for vertical separators is either 2.0-3.0 if gas flow-rate determines the size of the vessel or 2.0-6.0 if the liquid flow-rate determines the separator size.

According to Walas (1990), separator aspect ratio should be in the range of 2.5 to 5.0 based on the operating pressure of separator, as presented in Table 2-1. This approach was confirmed by Svrcek and Monnery (1993).[1]

Arnold and Stewart (2008) suggested applying a range of 1.0-4.0, supporting the empirical observation that liquid droplets may be re-entrained by the gas stream if the aspect ratio of the separator is higher than 4.0. However, as discussed by Grødal and Realff (1999), it should be noted that the re-entrainment probability (and, hence, the upper limit of the aspect ratio) is a strong function of the fluid physical properties which vary significantly from oilfield to oilfield.

Table 2-1. The aspect ratios proposed by Walas (1990) for multiphase separators.

P (kPa)	0-1700	1700-3400	>3400
Separator Aspect Ratio	3	4	5

[1] Please refer to Appendix A for more details.

2.1.2 *Two-Phase Separators*

In this subsection, the systematic guidelines which are based on droplet settling theory are reviewed. Then, the most common heuristics used for designing two-phase separators are presented.

2.1.2.1 Droplet Settling Theory for Vapor-Liquid Separation

In two-phase separators droplets of liquid are removed from a gas phase. Separation of liquid droplets from gas phase is described by settling theory in that the terminal velocity of a small droplet in a gas phase is governed by Equation 2-1 (Green and Perry, 2008):

$$U_d = \sqrt{\frac{4\times10^{-6} g d_p (\rho_d - \rho_c)}{3 C_D \rho_c}} \qquad \textbf{(2-1)}$$

where, U_d is settling velocity in *m/s*, g is gravity acceleration in m/s^2, d_p is droplet diameter in μm, ρ_d and ρ_c are dispersed phase (liquid) and continuous phase (vapor) densities (respectively) in kg/m^3, and C_D is drag coefficient which can be calculated via Equations 2-2 and 2-3 (Monnery and Svrcek, 2000) and is based on Gas Processors Suppliers' Association (GPSA) approach:

$$C_D = \frac{5.0074}{\ln(x)} + \frac{40.927}{\sqrt{x}} + \frac{44.07}{x} \qquad \textbf{(2-2)}$$

$$x = \frac{3.35\times10^{-9} \rho_c (\rho_d - \rho_c) d_p^{\,3}}{\mu_c^{\,2}} \qquad \textbf{(2-3)}$$

where μ_c is viscosity of continuous phase (vapor) in *Pa.s*.

Walas (1990) reported that sprays in process equipment usually are greater than 20 μm, mostly greater than 10 μm. His recommended droplet size for design purposes is 200 μm. Wu

(1990) proposed that for steady state operations, separators can be designed based on a droplet size of 250 μm.

Arnold and Stewart (2008), from field experience, showed that if droplets smaller than around 140 μm are removed in the gravity separation zone of the separator, the demister will not become flooded and will remove those droplets between 10 and 140 μm in diameter. So, their recommended droplet size for designing a two-phase separator is 140 μm. They also suggested using a higher value of 500 μm for designing gas scrubbers and a higher value in the range of 300-500 μm for designing flare or vent scrubbers.

In the separator design literature, Newton's equation (Equation 2-1) is also represented as the Sauders-Brown equation, Equation 2-4:

$$U_d = K\sqrt{\frac{\rho_d - \rho_c}{\rho_c}} \qquad \textbf{(2-4)}$$

where K is the settling velocity coefficient in *m/s*. Gerunda (1981) indicated that K ranges from 0.03 to 0.107 *m/s*, and for satisfactory designs K can be assumed to be 0.07 *m/s*.

For separators equipped with wire mesh demisters, Branan (1983) developed an empirical correlation for K coefficient, Equation 2-5:

$$K = \frac{0.0802}{x^{1.294} + 0.573} - 0.0022; \qquad 0.04 \le x \le 6.0 \qquad \textbf{(2-5)}$$

where x is a function of the flow-rates and phase densities as per Equation 2-6:

$$x = \frac{Q_d}{Q_c}\sqrt{\frac{\rho_d}{\rho_c}} \qquad \textbf{(2-6)}$$

where Q_d and Q_c are dispersed phase (liquid) and continuous phase (vapor) flow-rates in m^3/s, respectively.

Smith (1987) indicated that K coefficient should be considered as a function of many parameters such as separator orientation and aspect ratio (the most important factor), the design of the internals, feed conditions, operating pressure, foaming tendency of the fluids, and the degree of separation required. He suggested using a *K* coefficient ranging from 0.03 to 0.051 *m/s* (typically, 0.051 *m/s*) for vertical separators and a *K* coefficient ranging from 0.106 to 0.215 *m/s* (typically, 0.152 *m/s*) for horizontal separators. He also recommended using conservative values for unknown operating conditions or non-ideal cases such as slugging flow, high pressure, and excessive platform vibration.

For vertical separators, Walas (1990) proposed a *K* value of 0.0427 *m/s* for "no demister" cases and the *K* coefficients as shown in Table 2-2 for "with demister" cases. He pointed out that if the wire mesh pad is installed in a vertical or inclined position, the values of *K* should be 2/3 of the values reported in Table 2-2.

Evans (1974) and Walas (1990) recommended the use of higher values by a factor of 1.25 for horizontal separators. The York Mist Eliminator Company recommended that the *K* coefficients be based on the operating pressure. The values were regressed by Svrcek and Monnery (1993) and are presented in Table 2-3.

Table 2-2. The *K* coefficients proposed by Walas (1990) for multiphase separators equipped with stainless steel mesh pads.

Efficiency	Density (kg/m^3)	Special Surface (m^2/m^3)	K (m/s)	
			Under Pressure	Vacuum
Low (99.0%)	80-112	213	0.122	0.061-0.082
Standard (99.5%)	144	279	0.107	0.061-0.082
High (99.9%)	192	377	0.107	0.061-0.082
Very High (>99.9%)	208-224	394	0.076	0.061-0.082

Table 2-3. The *K* coefficients proposed by York Mist Eliminator Company and regressed by Svrcek and Monnery (1993) for multiphase separators equipped with stainless steel mesh pads.

Range of Absolute Pressure (kPa)	K (m/s)
6.7-101.3	$0.02843+1.28\times10^{-4}P+0.01402\ln(P)$
101.3-276	0.1067
276-37911	$0.1445-0.007\ln(P)$

GPSA guidelines also recommended that K coefficients should be estimated based on the operating pressure (Gas Processors Suppliers Association, 1998). The values proposed for vertical separators equipped with demister have been fitted and the result is shown as Equation 2-7:

$$K = 0.11[1.07 - 0.074\ln(1+0.01P)] \qquad \textbf{(2-7)}$$

where P is operating pressure in *kPa*. A factor of 1.25 and 0.5 should be multiplied by Equation 2-7 to calculate K coefficient for horizontal separators and separators without a demister pad, respectively. For vacuum services a value of K = 0.061 *m/s* for both vessel orientations was recommended.

2.1.2.2 <u>Heuristics</u>

2.1.2.2.1 Liquid Retention Time

Evans (1974) recommended a liquid retention (or residence) time of 2-5 *min* to process reasonable variations in fluid flow rates. For many cases, a liquid retention time of 5-10 *min* in a separator operating half-full of liquid is adequate (Walas, 1990). Gerunda (1981) recommended the use of a vertical separator if a short liquid retention time is acceptable.

Svrcek and Monnery (1993) recommended using a systematic approach for estimation of the required liquid retention in that the low liquid level is selected from the operating pressure and the vessel diameter, and then holdup[2] and surge time[3] are estimated based on the process type, personnel and instrumentation quality. Typical values for holdup and surge time are 2-10 *min* and 1-5 min, respectively (Svrcek and Monnery, 1993).

For smooth operation and control, Sinnott (1997) recommended a typical value of 10 *min* as the liquid retention time. Smith (1987) and Arnold and Stewart (2008) proposed that retention

---

[2] Holdup, considered for smooth and safe operation of downstream facilities, is defined as the time it takes to reduce the liquid level from normal to low liquid level by closing inlet flow while maintaining a normal outlet flow.

[3] Surge time, considered for handling upstream or downstream variations, is defined as the time it takes to increase the liquid level from normal to high liquid level by closing outlet flow while maintaining a normal inlet flow.

times of 30 *s* to 3 *min* are generally sufficient for most applications and can also satisfactorily process "slugs" or "heads" of well fluids. However, Smith (1987) pointed out that for satisfactory separation of foaming oil, the retention time should be increased to 5-20 *min* based on the foam stability.

2.1.2.2.2 Overall Aspects of Separator Geometry

Watkins (1967) developed some heuristics that account for the influence of instrument quality, labor quality, operating characteristics, and level control quality in specifying liquid holdup. These heuristics are summarized and illustrated in Figure 2-1. This guideline suggested that if the calculated aspect ratio is less than 3, the length is increased making the ratio equal to 3, and if the aspect ratio in a vertical separator is more than 5, a horizontal separator is instead specified.

Smith (1987) recommended that for a vertical separator, liquid depth above the oil outlet should be from one to three times the vessel diameter. In a horizontal separator, liquid phase should have a depth of 0.08 *m* to 60-70% of the separator cross-sectional area, with a typical value of one-third of the vessel diameter.

Walas (1990) recommended some key dimensions for designing separators equipped with mesh pad demisters which are illustrated in Figure 2-2. For vertical separators, Sinnott (1997) suggested a vapor-liquid disengagement height equal to the vessel diameter.

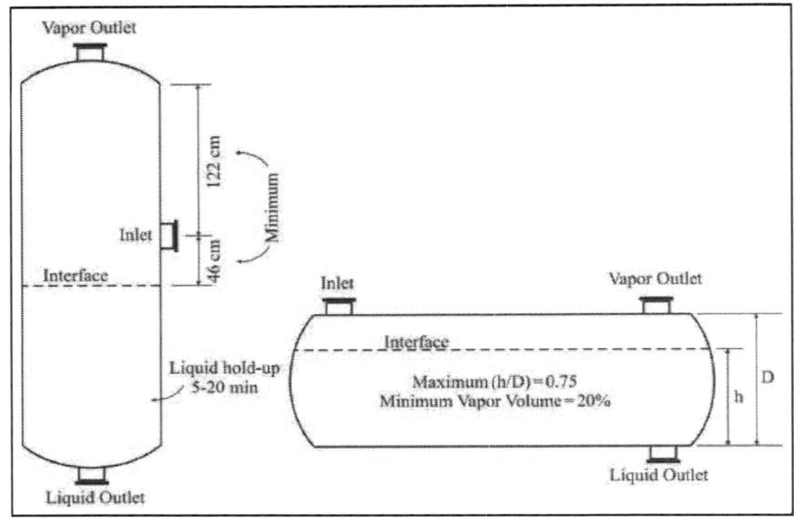

Figure 2-1. Design Heuristics Proposed by Watkins (1967) for Two-Phase Separators.

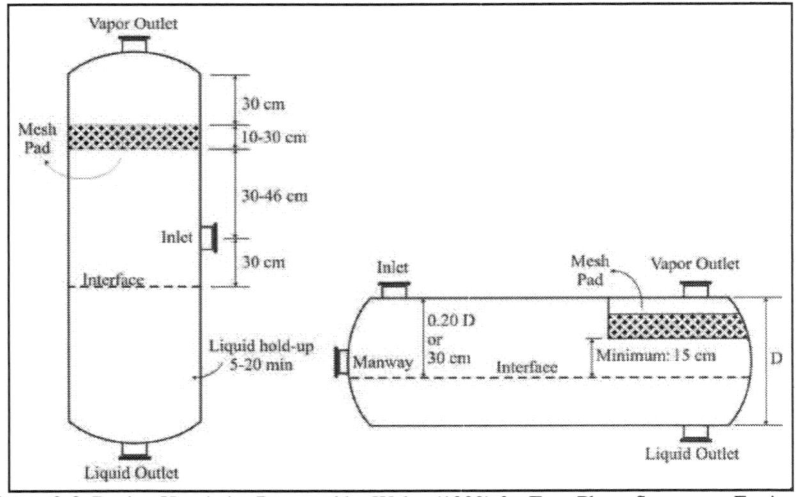

Figure 2-2. Design Heuristics Proposed by Walas (1990) for Two-Phase Separators Equipped with Wire Mesh Demisters.

2.1.3 *Three-Phase Separators*

In three-phase separators, both vapor-liquid and liquid-liquid separations must be considered in the separator sizing calculations. The compartment for vapor-liquid separation in a three-phase separator is designed based on the settling theory as presented for two-phase separators. Therefore, vapor-liquid separation need not be discussed in this section. However, the settling theory used for liquid-liquid separation and the most common relevant heuristics are presented in this section. As noted previously, the algorithmic procedure proposed by Monnery and Svrcek (1994) for designing three-phase separators have also been presented in Appendix A, providing additional details for the classic design methods.

2.1.3.1 Droplet Settling Theory for Liquid-Liquid Separation

Stokes' law is used to determine the separation (settling or rising) velocity of the droplets (Green and Perry, 2008):

$$U_d = \frac{d_p^2 g(\rho_d - \rho_c)}{18\times10^{12}\mu_c} \qquad \textbf{(2-8)}$$

where, U_d is settling (or rising) velocity in *m/s*, d_p is droplet diameter in μm, g is gravity acceleration in m/s^2, ρ_d and ρ_c are dispersed phase and continuous phase densities (respectively) in kg/m^3, μ_c is continuous phase viscosity in *Pa.s*. Abernathy (1993) and Monnery and Svrcek (1994) proposed that a maximum practical separation velocity of 0.00423 *m/s* be used as an upper limit in Equation 2-8. The accepted droplet size of 150 μm has been used as a standard in the API design method (Walas, 1990; Hooper, 1997).

Recent papers suggest assuming a larger droplet size for design purposes. Moreover, since the primary purpose of three-phase separation is to prepare the light liquid (usually oil) for further treating, separator design methods have usually addressed only water droplet removal from the oil phase. Field experience with separators sized in this way indicates that the oil content in the separated water phase is very low ranging from a few hundred to 2000 *mg/L* (Arnold and Stewart, 2008). For the rare case of operating very high-viscous water, Arnold and

Stewart (2008) recommended that an oil droplet size of 200 μm may be assumed for satisfactory separation of oil droplets from water phase. Lyons and Plisga (2005) and Arnold and Stewart (2008) recommended using a water droplet size of 500 μm for satisfactory separation of water droplets from the oil phase. An effluent emulsion to be treated by downstream equipment is expected to contain less than 5-10% water (Arnold and Stewart, 2008). For heavy crude oil cases, Arnold and Stewart (2008) recommended a water droplet size of 1000 μm, with the corresponding emulsion being expected to contain as much as 20-30% water.

2.1.3.2 Heuristics

2.1.3.2.1 Liquid Retention Time

Smith (1987) found that the operating liquid retention time in oil production separators varies from 20 *s* to 1-2 *h*. However, a typical retention time in three-phase separators is from 2 to 10 *min*, with 2-4 *min* being the norm (Smith, 1987). For cases where emulsions are not likely to form, Sinnott (1997) recommended a retention time of 5 to 10 *min* to be sufficient, while Perry and Green (1999) reported that in the petroleum industry, oversized separators with liquid retention times of up to 1.0 *h* have frequently been used, which in most cases are excessive and expensive. Arnold and Stewart (2008) recommended using a retention time ranging from 3 *min* to 30 *min* depending on laboratory or field data. For cases such empirical data are not available, they suggested using an oil retention time of 5 *min* to 10 *min* (proportional to oil gravity or viscosity) and a water retention time of 10 *min*.

2.1.3.2.2 Overall Aspects of Separator Geometry

The liquid level in a horizontal separator for three-phase operation is normally assumed to be at the horizontal centerline of the vessel (half full operation) to maximize the gas-liquid interface area (Smith, 1987; Arnold and Stewart, 2008). However, the liquid level can vary from 0.20-0.27 *m* up to 80-90% of the cross-sectional area of the vessel (Smith, 1987).

2.2 CFD-Based Studies of Multiphase Separators

Most of the CFD-based studies of multiphase separators have been performed by two research groups: NATCO and SINTEF. Both groups have developed CFD models to study the separation performance of large-scale separators. These studies are reviewed in the following two subsections. The other CFD-based studies will be presented in a separate subsection.

2.2.1 *Contribution of NATCO Group*

The biggest contribution to CFD based study of the separator performance is provided by National Tank Company (NATCO). The relevant studies of NATCO research group are reviewed in this subsection.

Frankiewicz et al. (2001) reviewed the effects of some design options, such as inlet distributors and distributing baffles, on reducing the size and weight of separation trains while maintaining or improving their performance. CFD results in terms of fluid streamlines have been presented for oil and water separation inside a three-phase separator with and without flow-distributing baffles. Moreover, contours of density have been presented for oil and gas separation inside a vertical two-phase separator. The sensitivity of operating quality of the installed vortex cluster to the inlet flow rate has been demonstrated. This paper did not include any information on the developed CFD models.

Frankiewicz and Lee (2002) used CFD simulations to study the flow pattern in two and three-phase oilfield separators. The influence of inlet nozzle configuration, flow distributors, perforated plates, and outlet nozzles were studied. It was realized that CFD simulations can be used in designing separator internals such as perforated plates in order to establish a reasonable flow distribution while minimizing liquid sloshing for offshore applications. The well-known $k-\varepsilon$ turbulence model and the Volume of Fluid (VOF) multiphase model of Fluent 6.0 were used, and the grid system was generated in the Gambit environment (Frankiewicz and Lee, 2002). The CFD results in terms of streamlines and velocity vectors for the fluid flows were considered as the criteria in modifying the separator internals. The inlet nozzle, the inlet momentum breaker, perforated plates, weir or bucket plates, and outlet nozzles were taken as the key components affecting the fluid streamlines throughout the CFD studies.

The single perforated plate installed in the separator was modeled through the porous media model of Fluent 6.0. As also confirmed by experimental tests, the CFD results predicted a non-plug (short-circuiting) flow regime downstream of the perforated plate (Frankiewicz and Lee, 2002). It was noted that fluids would always follow the path of least resistance to the separator outlet. This behavior resulted in a significant loss in the effective liquid retention time. To prevent this problem, CFD studies confirmed that a second perforated plate just upstream of the outlet nozzle was required. CFD results also predicted a high velocity profile for liquids flowing under perforated plates when an open area was retained to allow for sand migration. Such flow profiles not only favored water short-circuiting, but also increased the probability of oil and water mixing, particularly if the oil-water interface were too close to the lower opening area. The position of perforated plates was determined by experience and practical considerations such as location and configuration of the inlet nozzle, position of the weir and water outlet nozzle, and the vessel dimensions. It was emphasized that a fully symmetric setting of the internal baffles was not effective.

CFD simulations of Frankiewicz and Lee (2002) were validated by laboratory tests and showed that the standard box distributor tended to bypass a significant fraction of emulsion around the electrostatic grid, and the device should be replaced by a shrouded pipe distributor. Revamping the operating separator with composite plate electrodes and the shrouded pipe distributors led to an increased capacity of some 67%, due mainly to improved flow distribution, i.e. more plug flow regime.

The primary goal of the CFD study performed by Lee et al. (2004) was to evaluate the design of internals for a three-phase separator that would mitigate sloshing of liquid phases due to offshore wave motions. The separator of interest was installed as part of a Floating Production Storage and Offloading (FPSO) facility. The internals simulated included perforated plate baffles, combo plate (comprised of both solid and perforated portions), and weir were modeled as either porous media or solid wall in the CFD simulations.

Three case studies were performed. The first two were completed with a "no flow" assumption since fluid flow practically had only a small effect on the sloshing motion of interfaces. Fluid profiles obtained for the first two case studies demonstrated that the separator would not operate properly because the water phase would spill over the oil weir if the installed baffles were excluded from the separator. To prevent the water phase from being pulled up toward the oil weir, a perforated baffle was designed and placed near the oil weir, and the open areas of two preceding baffles were decreased. Although the position of oil weir was fixed, its configuration was modified based on the results of CFD simulations. These improvements led to a reasonable control of oil-water interface, eliminating the water spillover problem of the second case study. Finally, as a third case study, fluid flows of oil and water phases were taken into account while evaluating the general performance of the installed baffles. In this case study, assuming that there would be a minor impact from gas phase flow on the oil-water interface, gas flow was ignored. The results of CFD simulations indicated that the designed baffle arrangement, which was the same as the second case study, would mitigate the sloshing problems and keep the water phase from spilling over the weir although both sources of turbulence, wave motions and liquid flows through the vessel, were present. The field data showed very low levels of impurities in the separated oil and water phases, thus confirming both the validity of the CFD simulation studies and successful operation of the separator. Unfortunately, the details of the CFD simulations and the obtained solutions were not presented in the paper.

Lu et al. (2007) evaluated the effectiveness of perforated plate baffles for improving the separation performance of a FWKO drum/separator. The FWKO separator of interest was a horizontal vessel with diameter of 5.7912 m and length of 19.812 m. The normal liquid level and oil-water interface level were 4.319 m and 2.858 m, respectively. Operating and physical parameters including pressure, temperature, flow rates, densities, and viscosities were also given. The separator design was based on providing a residence time of 18.8 minutes for the oil phase and a residence time of 18.4 minutes for the water phase. Considering the symmetric geometry and configurations of the separator, only half of the vessel was modeled. This simplifying assumption was not reasonable because of the large deviation from plug flow regime obtained for fluid flows inside the separator. The grid system was generated in Gambit environment leading to some 245000 computational cells. The quality of the produced mesh system was not verified in the study. Based on the vessel dimensions, it seems reasonable that many more grid cells should be generated, providing a fine mesh system. The $k-\varepsilon$ turbulence model and the Mixture multiphase model of Fluent 6.2 were used. Multiphase modeling was not based on CFD simulation guidelines, but on a balance between model capabilities and available computational resources.

The two perforated plate baffles, designed to improve the flow distribution in a FWKO separator, with specific configurations were modeled as some porous jump boundaries. Although details of modeling were not provided in the paper, it seems that the porous jump model (Fluent 6.3 User's Guide, 2006) was selected because of its ease of implementation. However, the model recommended for flow distributing baffles is the porous media model in spite of the more difficult solution convergence involved.

The CFD results in terms of velocity contours of fluid flows were presented to visually confirm that by installing the perforated plate baffles, flow distribution was improved as the large flow circulations were broken into small ones. Furthermore, the Euler-Lagrange approach (probably through Discrete Phase Model (DPM) of Fluent 6.2) was then used to track the trajectories of fluid particles. The mean residence time of fluid particles were calculated and used as the criteria for evaluating separation performance of the perforated plate baffles. As the result of the installation of distributing baffles, the mean residence time increased from 630 s to 980 s for the water phase and from 520 s to 745 s for the oil phase. The volumetric utilization, defined as the ratio of actual residence time to theoretical residence time, also increased from 46% to 66% for the oil phase and from 57% to 89% for the water phase.

Lee et al. (2009) presented some engineering suggestions and corresponding CFD based verifications performed to revamp the phase separation inefficiencies experienced in a major oil production facility. One of the separators of interest was operating at a high pressure of 400 kPa, and the other was operating at a low pressure of 100 kPa. The high-pressure separator of interest was a horizontal vessel with diameter of 4 m and length of 10.4 m, and the low-pressure separator was a horizontal vessel with diameter of 4.375 m and length of 13.5 m. All the operating and physical parameters except for configuration of distributing baffles were also provided for both separators. Separator design had been based on providing a liquid (both oil and water) residence time of approximately 5.5 minutes in the high-pressure separator and a liquid residence time of approximately 4 minutes in the low-pressure separator. "Debottlenecking" studies led to some suggestions for the weir height, liquid levels, and configuration and position of distribution baffles.

CFD simulations were performed to evaluate the overall improvements resulting from modifications to the vessel internals and settings. Generation of grid system was completed in Gambit environment (Gambit 2.4.6, 2006), and the CFD simulations were performed using the Fluent 6.3.26 software. Although details of modeling have not been provided in the paper, it appears that almost the same strategies as proposed in the previous study of Lee et al. (2007) had been used. Again, the volumetric utilization has been considered as the criteria in evaluating separation performance of the modified internals and settings. As the total result of all modifications, the volumetric utilization in the high-pressure separator increased from 82% to 84% for the oil phase and from 55% to 75% for the water phase. In the case of the low-pressure separator, the volumetric utilization increased from 87% to 94% for the oil phase and from 52% to 95% for the water phase. Therefore, the CFD simulation results showed that the separator modifications mainly influenced the water phase. Also, the CFD results for the fluid flow streamlines were presented to visually confirm that by implementing the suggested modifications, flow distribution was improved as the large flow circulations were broken into small weak ones.

2.2.2 *Contribution of SINTEF Group*

SINTEF group provides the second biggest contribution to CFD based study of the separator performance. The title of this Norwegian research group stands for "Stiftelsen for industriell og teknisk forskning" which means: Foundation for Scientific and Industrial Research. The relevant studies of SINTEF group are reviewed in this subsection.

Hansen et al. (1991) introduced an under development computer code, Flow Simulator for Separators (FLOSS), aimed for simulation of fluid flow behavior inside the phase separators. A cubical transparent model with dimensions of 0.46 $m \times 0.46\ m \times 1.83\ m$ was also developed to validate the results of the separator simulator. The experimental separator model was equipped with a spherical deflector baffle as the momentum breaker and one flow distributing perforated baffle. The configuration and all geometrical specifications of the momentum breaker and perforated baffle have been provided. The experiments were performed using air, water, and a special oil mixture as the fluids, and the fluid flow behavior in inlet zone and in liquid gravity separation zone, in which the oil-water separation occurs, was of specific interest. Using a laser-Doppler flow-meter and by tracking an injected tracer, velocity and residence time distribution in the zones of interest were measured to produce the experimental data that was compared with the results of the FLOSS simulations. The experiments showed that the fluid flows in the gravity separation zone deviated from uniform plug flow pattern, and it was more than likely that some recirculation regions existed.

Two multiphase models were implemented to simulate the various features of phase separation:

1. Two-Fluid Model; based on the comments given in the paper, it appears that this model was a reduced form of the more general multiphase model of "Volume of Fluid"[4] arranged for two immiscible phase separation simulation. This model was used for simulation of vapor-liquid separation in the inlet zone of the separator.
2. Drift-Flux Model; similarly, it appears that this model was a specific version of the more general "Mixture" multiphase model which was used for simulation of separation of oil and water droplets dispersed in the liquid phases.

Finally, it has been shown that the simulation results in terms of velocity profiles and actual liquid residence time were in good agreement with experimental data. For example, average residence time of the tracer was less than the theoretical value in both the numerical simulation and the experimental study. Unfortunately, details of the CFD modeling are missing from this paper, and the multiphase CFD model has been limited to two-phase simulation.

[4] Please refer to section 3.1.2 for more details.

Hansen et al. (1993) presented the simulation results of the developed CFD code, FLOSS, for an industrial scale three-phase separator. The separator of interest, with diameter of 3.33 *m* and length of 16.30 *m*, was the first stage of the three-stage-dual-train production process installed on Gullfaks-A offshore platform. The inlet three-phase fluid flow entered the vessel as a high momentum jet striking a spherical deflector baffle as the momentum breaker. The upper part of the vessel was equipped with internals, including flow distribution baffles and demister, to enhance the separation of liquid droplets from gas.

During the first four years of operation, some operating problems and separation inefficiencies, such as emulsion problems, and water level control failure with increasing produced water, were experienced. The CFD studies were performed by Hansen et al. (1993) in order to develop a deep understanding of the complex three-phase separation process taking place inside the separator. In their modeling effort, Hansen et al. (1993) focused on two zones: the inlet and momentum breaker zone, and the bulk liquid flow zone. The grid systems used for the numerical simulations of the inlet zone and the bulk liquid zone were $11\times8\times15$ and $23\times4\times5$, respectively. The inlet section of the separator in which all three phases are present was modeled as a two-phase gas-liquid flow, and the results did provide the boundary conditions for the distributed velocity field in the liquid pool. The approach was taken to simplify the complicated features of the problem. The other important simplifying assumption, in both zones, was to assume a symmetrical fluid flow profiles around the vertical plane in the middle of the separator, thus, only half of each zone volume was modeled.

Hansen et al. (1993) proposed modifications to the separator based on the CFD predictions of rotational flow regimes established between any two baffles. Testing of the modified separator indicated improved data in terms of the water level control and the produced oil quality.

Because of the problem scale and importance, and also since almost all the operating and physical parameters required for CFD simulations have been provided by Hansen et al. (1993), this significant case study was selected for comprehensive CFD studies. Further comments on the performed CFD simulations and results, as well as assessment of the original work approach and results are presented in Chapter Four.

Hansen et al. (1995) addressed some practical aspects of the phase separation inside the multiphase separators which led to using drift-flux and two-fluid models as the multiphase models in their CFD simulator, FLOSS. Due to difficulties in trying to extend the single-phase CFD model to multiphase version, the computer code development was confined to two-phase fluid flow simulation.

Potential capabilities of the developed CFD code, FLOSS, in providing fluid flow regimes and velocity vector profiles in the various zones of the phase separators were demonstrated. The typical results were given for the inlet zone of a three-phase separator, the first stage separator of Gullfaks-A platform, and the bulk liquid zone of a test separator. Details of the CFD modeling have not been presented in their paper (Hansen et al., 1995).

Hansen (2001) reported some overall features of experimental and CFD simulation studies for oil-water distributions. First, various correlations proposed for estimation of the liquid viscosity in oil-water mixtures were reviewed. This study indicated that although the correlations agree well for low-water mixtures, they produce totally different values for mixtures with higher water-cuts. Then, Hansen (2001) described the bottle test in which batch separation of oil-water mixture is tested. The performed experiments showed that the oil-water mixture behaves very similar to a homogeneous fluid flow. Thus, the drift-flux model of FLOW-3D software was used for simulation of this batch separation. The coalescence model of this software was further developed and settling of water droplets (with mean diameter size of 160 μm) in a 20% water-cut mixture was simulated and the simulated settling time approached the value predicted by the hindered Stokes' equation discussed in Kumar and Hartland (1985).

Two-phase fluid flow in the liquid zone of a horizontal separator, with diameter of 1.95 m and length of 6.0 m, was next simulated. For this purpose, the VOF model of FLOW-3D software was used. The CFD simulation results in terms of velocity profiles demonstrated an inhomogeneous flow regime in the liquid zone. Some backflow and recirculation zones were predicted for the gas-liquid interface and distribution baffle vicinities. Unfortunately, the details of the developed CFD models were not provided in the paper.

2.2.3 *Miscellaneous Studies*

Fewel and Kean (1992) provided a review of gas-liquid separation issues with a focus on vane-type demisters. After addressing the advantages of vane-type demisters, such as their higher capacity compared to mesh pads, the well-known relationships and approaches for evaluating separation efficiency, pressure drop, and capacity of vane-type demisters were presented. It was noted that evaluating the capacity of a separation device is the most important stage in design process, and this significant issue is frequently overestimated because of the errors involved in scale-up of experimental results. Moreover, physical properties of fluids, such as their density, viscosity, and surface tension, in the oilfield operation are typically significantly different from those of laboratory test fluids. Therefore, both the flow pattern of gas phase and droplet size distribution of liquid phase are also different. To overcome this shortcoming, it has been suggested that field experience be combined with CFD simulation analyses to obtain a good estimation of the vane-type demister performance.

The authors then briefly described how porous cells can be defined in a CFD model to simulate some separation devices such as mesh pads, vanes, and filters. Fewel and Kean (1992) have obtained converged solution of the CFD model to check if determined allowable velocity is exceeded in any areas of the vane. If so, the number of vanes may be increased or some perforated baffles may be added to the design in order to prevent liquid droplet re-entrainment and improve the separation efficiency of the vane-type demister. CFD simulations design can also check the modified internals to make sure that gas-liquid separation is effectively achieved.

Fewel and Kean (1992) have pointed out that the CFD analysis of separator internals is very similar to a physical test because laboratory tests performed on various arrangements usually match CFD results remarkably well. Therefore, it has been concluded that CFD studies can provide accurate gas capacity predictions. For example, in a case study (Fewel and Kean, 1992), velocity vectors of the CFD simulation matched physical measurements within 5% in a redesigned vane-type demister, and successful operation of the demister led to exceptional separation performance of the separator. Full gas capacity of the separator was achieved and no trace of liquid hydrocarbon carryover by gas phase was observed.

Wilkinson and Waldie (1994) used two pilot plant scale transparent horizontal separators; one was two-dimensional and the other was three-dimensional. They also developed their corresponding CFD models to study the flow pattern of oil-water separation phenomenon. The two-dimensional separator was rectangular with dimensions of 0.23 $m \times 0.25$ $m \times 0.875$ m, and the three-dimensional separator was cylindrical with diameter of 1.0 m and length of 3.77 m. In both separators, geometrical specifications of inlet, water outlet, weir plate, oil outlet, and liquid levels have been given. In order to measure fluid velocity and particle size distribution in the experimental models, Laser Doppler Anemometry (LDA) and Phase Doppler Analysis (PDA) techniques were used.

Experiments were conducted using dispersed oil in a oil-water mixture (with volumetric fractions of 0.01 to 0.10%) at superficial velocities from 0.005 *m/s* to 0.012 *m/s* for the two-dimensional model and oil-water mixtures at superficial velocities from 0.005 *m/s* to 0.010 *m/s* for the three-dimensional model. Although, from a practical point of view, using oil-water mixtures with higher oil concentrations is more realistic for field separator operations, this led to some technical problems in PDA measurements. Therefore, the measureable fluid flow systems were studied.

The $k-\varepsilon$ turbulence model and the Discrete Phase Model (DPM) of the commercial CFD software, Fluent 2.99, were used for the CFD simulations. The grid systems used for numerical simulations of the two and three-dimensional models were 25×75 and 11×11×41, respectively. In the case of three-dimensional model, flow was considered to be symmetrical around the vertical plane in the middle of the separator, thus, only half of the volume was modeled.

The CFD results in terms of velocity profiles and particle size distribution in the inlet vicinity were in reasonable agreement with experimental data in the two-dimensional model. However, in contrast with the CFD results, the PDA measurements detected a significant presence of oil droplets in the inlet zone of the separator. This discrepancy was justified by the fact that in CFD simulations, an ideal steady state operation of separator was assumed even though the practical operating conditions were fully dynamic.

Furthermore, there were substantial differences between measured flow patterns and CFD simulations in the three-dimensional model. This shortcoming was mainly associated with some limitations in computational resources which restricted the number of defined grid cells. Actually, the minimum grid cell size which could be used was restricted by the maximum number of cells in the grid system of 20001. Therefore, a relatively coarse grid system with the average cell size of 0.036 $m \times 0.045\ m \times 0.092\ m$ had to be used for CFD simulation of three-dimensional model.

The authors concluded that although the results were not necessarily representative for practical oil-water separators, they did demonstrate some potential problems while applying CFD simulations or measurement techniques to multiphase separation phenomenon (Wilkinson and Waldie, 1994).

Hallanger et al. (1996) developed a CFD model for a three-phase separator by extension of Two-Fluid model. The phases of interest were free gas, oil containing dispersed water, and free water. The water droplets dispersed in the oil phase were assumed to be spherical and obey the drag law for solid particles. The Mixture model was used for modeling the oil phase. The water droplets were distributed among different classes based on their diameter size, and a momentum equation for the mixture phase together with continuity equations for each class were solved. Interactions between dispersed droplets such as coalescence and breakup were neglected. The pressure correction approach, as proposed through the Semi-Implicit Method for Pressure Linked Equations (SIMPLE), with some adjustments for the mixture phase was used to obtain the numerical solution of the system.

The model was used to simulate the first-stage separator, with diameter of 3.15 m and length of 13.1 m. The other geometrical aspects of the separator are missing from the paper. The separator was equipped with a deflector baffle, two perforated baffles, a demister, and a weir plate. Assuming symmetrical flow profiles, only half of the vessel was modeled. The grid system used for numerical simulation was $10 \times 20 \times 49$, corresponding to the average cell size of 0.1575 $m \times 0.1575\ m \times 0.27\ m$. For definition of dispersed water droplets, an average diameter of 0.25 mm with spread parameter of 3.0 was used, and droplets were distributed in seven particle classes.

The velocity profiles for gas, oil, and water phase flows have been presented and discussed. CFD results also indicated that most of the smallest water droplets would remain in the oil phase while almost all of the largest droplets would join the free water phase. The CFD results in terms of concentrations of water droplets in the oil outlet versus oil residence time were in reasonable agreement with experimental measurements. The largest deviation from experimental data occurred for a long oil residence time of around 130 *s*.

Wilkinson et al. (2000) performed small scale experiments and some CFD simulations to investigate the flow of a single liquid phase in a two-dimensional model separator prior to switching to larger three-dimensional two-phase (oil and water) models. The best separation was achieved when the flow of liquids was close to plug flow in the separator without regions of re-circulating flow. The aim was to design simple internal fittings to achieve a plug flow velocity distribution in the separator without inducing extra dispersion of the phases.

These results were compared with experimental data obtained from a larger three-dimensional cylindrical model with baffles using only water flows and a 20% by volume oil in water mixture. Flow distribution quality of a pair of baffle plates with various configurations were also measured in the three-dimensional cylindrical model.

Two pilot-plant-scale transparent horizontal separators and their corresponding CFD models were used to study the effect of distributing perforated baffles on fluid flow pattern in oil-water separation phenomenon. The two-dimensional separator was the same as in their previous study (Wilkinson and Waldie, 1994), but a smaller three-dimensional cylindrical separator with diameter of 0.60 *m* and length of 2.26 *m* was also used in this follow-up study. In both separators, geometrical specifications of inlet, water outlet, weir plate, oil outlet, and liquid levels have been provided in the paper.

Experiments were conducted using water at superficial velocity of 0.005 *m/s* for the two-dimensional model and either water or dispersed oil in water mixture (20% by volume) at superficial velocity of 0.011 *m/s* for the three-dimensional model. The velocities were chosen to provide a residence time of 100–200 *s* which is similar to oilfield production separators.

In order to measure fluid velocity in the experimental models, Laser Doppler Anemometry (LDA) and Phase Doppler Analysis (PDA) techniques were used. Significant velocity fluctuations, with typical standard deviation of 40% around the average velocity, were reported based on the PDA measurements in the two-dimensional model. In the three-dimensional model, flow distributing effects of perforated baffles were studied by installing a single baffle just downstream of the momentum breaker. Various perforated baffles with open area fractions of 5% to 20% were tested. The standard deviations of time averaged axial velocity measurements were used as criteria for evaluating the effectiveness of the flow distributing baffle. Since measurements in water only flows were consistent with those in dispersed oil in water flow, only two runs were performed using the oil-water mixture.

A two-dimensional CFD model was developed for each of the experimental separators using the $k-\varepsilon$ turbulence model of the CFD software, PHOENICS v.1.5. Without providing detailed experimental data, Wilkinson et al. (2000) concluded that the developed CFD model can provide a reasonably good simulation of the flow regime in the two-dimensional experimental separator. However, as indicated by the comparison between the results of CFD simulations and experimental data, two-dimensional CFD modeling of the three-dimensional separator was not so successful. Although the velocity profiles of the CFD simulation were partially the same as the experimental measurements, the optimum suggested by CFD was at a different value of baffle open area. There was also a significant difference in the magnitudes of standard deviation of velocities between the two-dimensional CFD model and the three-dimensional experimental model. However, the CFD model did predict that there should be an optimum value for baffle open area to maximize flow uniformity.

Swartzendruber et al. (2005) modeled a vertical two-phase separator by using the commercial CFD software, Fluent. The separator was equipped with a deflector baffle and a vane-type demister. The focus of the study was on the quality of gas flow distribution through the demister. Again, unfortunately, details of the developed CFD model are missing from the paper. CFD results such as the velocity profiles and fluid flow streamlines were presented. In order to mitigate the uneven flow distribution in the vane demister, as shown by the CFD simulations, the following changes were proposed:

1. The deflector baffle should be moved away from the inlet and installed parallel to vane demister.
2. A 90° elbow with turning vanes should be installed between the inlet and the deflector baffle.

It was also concluded that CFD simulations can be used as an effective design tool in identifying potential problems and modifying low-efficiency separator designs.

Newton et al. (2007) presented an introduction to the two modeling tools for multiphase separators, CFD and Visual Dynamic Modeling (VDM), and discussed some issues and limitations involved with implementing each approach. The VDM approach was described as a qualitative modeling tool which uses a scaled model of a separator to simulate the flow pattern inside the actual separator. No measurements are taken, and the model is used to visualize the behavior of multiphase flow through the separator.

CFD, as a modeling tool, was then introduced and its well-known multiphase modeling approaches, Euler-Lagrange approach and Euler-Euler approach, and the restrictions caused by available computational resources were addressed. To overcome the computational limitations, it has been suggested that the separator volume be divided into compartments; each including only one continuous phase and dispersed phases. Each compartment is simulated independently, while gas-liquid and liquid-liquid interfaces are defined as frictionless walls that will trap the dispersed droplets coming into contact with them. Therefore, the droplets are supposed to be eliminated from computational space after reaching these interfaces.

When specifying a dispersed phase in a CFD model, basic concepts such as using Rosin-Rammler equation for defining particle size distribution, and estimating the maximum stable particle size were presented. Then, some of the necessary steps for CFD simulation of multiphase separators using Euler-Euler approach, i.e. initializing the solution, "patching" volume fractions of the phases, setting under-relaxation factors based on convergence and stability trend of iterative solution, and periodic checking/adjusting the level of interfaces while iterations are proceeding, were addressed. The authors placed emphasis on careful implementation of the Euler-Euler approach due to its extremely complex calculations at each iteration.

In the last section of the paper, some important shortcomings of VDM approach for modeling multiphase separators were discussed. It has been pointed out that, because of some economic restrictions, VDM systems typically utilize two-phase flow of air and water at moderate pressures for visualizing multiphase flow of other materials inside the separators operating at higher pressures. It was also noted that in order to achieve an acceptable representation of the multiphase flow through the actual separator, the involved internals should be scaled accurately, and the Reynolds number of VDM flows should be equal or very close to the Reynolds number of the actual separator flows.

This paper presents some useful comments and remarks on basic concepts of CFD simulation of multiphase separators. However, it should be noted that independent simulation of separator compartments, as proposed by Newton et al. (2007) for simplification of the separator model, may lead to some abnormal results as will be discussed in more detail in Chapter Four.

2.3 Two Relevant Theses

Two academic theses have been developed to empirically study the performance of multiphase separators in oil production facilities:

1. "Gravity Separator Revamping" by Arntzen (2001)
2. "Experimental Characterization of High-Pressure Natural Gas Scrubbers" by Austrheim (2006)

In both of these research projects, pilot plant scale separators were used for the study of phase separation phenomenon. The Arntzen thesis focused on the study of liquid-liquid separation, while the Austrheim thesis focused on the study of vapor-liquid separation. These two significant research projects are of special interest to the present research project and are reviewed in detail in this section.

Arntzen (2001) focused on the mechanism of formation of oil-water interface (referred as dispersion layer in the thesis) and its mathematical modeling. A pilot-plant scale horizontal separator with diameter of 0.630 *m* and length of 2.80 *m* operating at room temperature and atmospheric pressure and equipped with an inlet cyclone and a flow distributer (diffuser) was used in the experiments. Different multiphase fluid systems, composed of mixtures of Exxsol oils with a mixture viscosity of 1.6 *mPa.s* and a crude oil with viscosity of 1.38 *mPa.s*, saline water, and air, were tested. The experimental data showed that the dispersion layer thickness increases almost linearly with dispersed phase concentration within the limited range investigated.

Coalescence and penetration of droplets through the interface were also analyzed by studying the water outlet quality. With water as the dispersed phase, the oil concentration at the water outlet increased by increasing pressure drop in the inlet valve, water flow-rate, and coalescence rate. With water as the continuous phase, the oil concentration at the water outlet increased by increasing pressure drop in the inlet valve, and water flow-rate. Furthermore, the data showed that the magnitude of pressure drop in the inlet valve would affect both dispersed and continuous phases, but the concentration of the dispersed phase and position of the inlet valve would affect only the continuous phase. Therefore, it was concluded that the dispersed phase quality might be affected only by the initial turbulence, while the continuous phase quality was affected by downstream coalescence.

To improve the pilot-plant-scale separator performance, three different mechanical alternatives, i.e. adjusting the position of inlet nozzle, removing a water-rich fraction of inlet flow before the flow diffuser, and increasing the coalescence area by installing some parallel horizontal baffles, were studied. The water outlet quality and oil-water interface thickness were used as the separation efficiency criteria. The performed experiments showed that the entrance position of the liquid was important, and the dispersion layer increased by injecting the liquid phase in a non-continuous liquid phase. These experimental results were also confirmed by tests on the large scale separators operating in the Gullfaks and Statfjord oilfields. The second mechanical alternative, removing a water-rich fraction of inlet flow upstream of the flow diffuser, was promising only for water dominant feeds. Finally, the use of coalescing baffles did not lead to a significant improvement in the separator performance.

The Austrheim (2006) study focused on the performance of gas scrubbers operating at low and high pressures. Three different vertical pilot-plant scale separators operating at various pressures were used:

1. A small scrubber with diameter of 0.389 *m* operating at ambient temperature and low pressures ranging from 200 *kPa* to 650 *kPa* was used. The multiphase fluid system was composed of the Exxsol D60 oil with viscosity of around 1.4 *mPa.s*, water and air.

2. A small scrubber with diameter of 0.150 m operating at ambient temperature and high pressures ranging from 2000 kPa to 9200 kPa was used. The involved two-phase fluid system was composed of the Exxsol D60 oil with viscosity of around 1.4 $mPa.s$ and nitrogen or, for some cases, synthetic natural gas containing methane, ethane, and pentane.
3. A large scrubber with diameter of 0.840 m operating at ambient temperature and high pressures ranging from 2800 kPa to 11300 kPa was used. A two-phase fluid system formed by processing a real natural gas was used.

All the pilot-plant separators were equipped with a vane-type inlet and a mesh pad demister. The experimental data showed that the separation performance of the scrubbers was excellent in non-flooded operating region of the demister. The flooding point occurred for the large scrubber at high pressure and processing a real natural gas. For this case, a significant liquid carryover was measured due to droplet re-entrainment (mainly) and some very small droplets. Therefore, it was recommended that more attention should be paid to understanding and evaluating re-entrainment mechanisms particularly while dealing with live hydrocarbon fluids at high pressures. In fact, separation efficiency of the scrubber generally decreased at high pressure operation with the live hydrocarbon fluid system.

Some significant discrepancies between the results of large scale scrubber and the small scale scrubber were reported. These discrepancies were associated with different distributions of the multiphase fluids in the cross-sectional area of the scrubbers. Thus, it was emphasized that in order to predict the performance of real natural gas scrubbers, the experimental tests should be carried out on large scale separators with oilfield fluids at relevant pressures.

Furthermore, droplets size distribution in a high-pressure scrubber was determined for the first time through experimental measurements. For this purpose, the laser scattering technology instrumented by Malvern Instruments Ltd. was used. The results indicated that in most cases the maximum droplet size was greater than 400 μm and the minimum droplet size was in the range of 1-10 μm.

2.4 Summary

The important relevant literature for the research project scope has been reviewed in this chapter. The heart of this chapter is review of the few CFD-based studies of multiphase separators. Two research groups, NATCO and SINTEF, and other researchers have provided a few papers on CFD-based study of multiphase separators. Although these studies have led to some beneficial modifications and useful scientific achievements, from an academic point of view, this part of "multiphase separator literature" is subjected to some significant shortcomings:

1. In the published documents, only the overall steps of CFD modeling have been provided and the details of developed CFD models are missing.
2. Generally, symmetrical fluid flow profiles have been assumed and only half of the separator volume has been modeled. Considering the fact that the plug flow regime assumption is not valid based on even these simplified CFD models, assuming symmetrical flow profiles is not realistic.
3. The quality of produced computational grid system generally has not been validated. In some studies, the computational restrictions have led to employing some coarse grid systems and, hence, dubious results.
4. In some studies, in order to reduce the required computational memory and effort, two-phase simulation of three-phase fluid flow has been performed. This approach evidently reduces the validity of produced CFD profiles.
5. Generally, some indirect factors such as the liquid retention time, the volumetric utilization, and the standard deviation of time averaged velocity have been used as the criteria for evaluating separation efficiencies. Although improving these factors can lead to more plug flow regimes inside separators, the actual separation efficiency is not necessarily improved. Therefore, the criteria are not real measures of separator efficiency.
6. CFD based modifications have generally been confined to the separator internals such as flow-distributing baffles. However, as emphasized by Lyons and Plisga (2005), optimizing the separator internals has only a minor effect on the separator performance,

and an inefficient/poorly designed separator cannot be significantly improved by optimizing its internals.

As will be explained through the following chapters of the book, the current research project will modify the above noted CFD simulation shortcomings with new comprehensive approaches and efficient strategies:

1. In this book, all details of developed CFD models will be provided and discussed.
2. Total volume of multiphase separators will be modeled.
3. The high quality of produced computational grid system will be validated.
4. All fluid phases present in a multiphase separator will be considered in CFD simulations, and more high quality details of the phase separation features will be provided.
5. Separation efficiency will be evaluated directly based on mass distribution of fine droplets as they are tracked by a suitable multiphase CFD model. Thus, the model will provide a realistic performance of the simulated separators.
6. Multiphase separators with all their internals will be simulated, and realistic modifications on an operating large-scale separator will be proposed. Moreover, the current research project will establish improved design criteria for existing design methods.

In this chapter, classic guidelines for design of multiphase separators have also been reviewed. In the classic methods, vapor-liquid and liquid-liquid separation compartments are designed based on droplet settling theory. Moreover, the retention time of liquid phase is selected based on empirical data or heuristics for establishing a safe and smooth operation of the separator and downstream equipment. Finally, two academic research projects, developed to empirically study the vapor-liquid and liquid-liquid separations in oil production systems, have been reviewed.

Chapter Three: Developed CFD Models

When one considers the scope of the research project, the size of the separators of interest, and the challenge involved with CFD simulation of multiphase fluid flows, a proven commercial CFD package should necessarily be selected for the development of the CFD models of the multiphase separators. In this research project, the commercial CFD package, Fluent 6.3.26, was selected for the model development. Note, this software package had been used successfully for simulation of large-scale multiphase separators by the NATCO group in numerous case studies.

This chapter will outline the steps required for the successful development of efficient CFD models using the Fluent software for simulating complex features of multiphase separation. In contrast with the published CFD based studies, which are mostly industrial assessments, all the details of the CFD modeling will be presented.

3.1 CFD Background

Before proceeding to the simulation case studies, the pertinent CFD concepts and some modeling strategies, particularly those involved in simulating multiphase separators, are introduced in this section.

3.1.1 *Approaches to Solving Equations for Fluid Flow*

In the modern CFD literature, mathematical representation of the governing equations for viscous fluid flow is addressed as the Navier-Stokes equations. In order to demonstrate their complex nature, these equations for a dynamic, three-dimensional, compressible, viscous flow are presented in the following:

- Continuity Equation (mass is conserved)

$$\frac{\partial \rho}{\partial t} + \nabla.(\rho V) = 0 \qquad \textbf{(3-1)}$$

- Momentum Equations (Newton's second law)

$$\frac{\partial (\rho u)}{\partial t} + \nabla.(\rho u V) = -\frac{\partial P}{\partial x} + \frac{\partial \tau_{xx}}{\partial x} + \frac{\partial \tau_{yx}}{\partial y} + \frac{\partial \tau_{zx}}{\partial z} + \rho b_x \qquad \textbf{(3-2)}$$

$$\frac{\partial(\rho v)}{\partial t}+\nabla.(\rho v V)=-\frac{\partial P}{\partial y}+\frac{\partial \tau_{xy}}{\partial x}+\frac{\partial \tau_{yy}}{\partial y}+\frac{\partial \tau_{zy}}{\partial z}+\rho b_y \qquad \textbf{(3-3)}$$

$$\frac{\partial(\rho w)}{\partial t}+\nabla.(\rho w V)=-\frac{\partial P}{\partial z}+\frac{\partial \tau_{xz}}{\partial x}+\frac{\partial \tau_{yz}}{\partial y}+\frac{\partial \tau_{zz}}{\partial z}+\rho b_z \qquad \textbf{(3-4)}$$

- Energy Equation (energy is conserved)

$$\frac{\partial}{\partial t}\left[\rho\left(U+\frac{V^2}{2}\right)\right]+\nabla.\left[\rho\left(U+\frac{V^2}{2}\right)V\right]=\rho\dot{q}+\frac{\partial}{\partial x}\left(k\frac{\partial T}{\partial x}\right)+\frac{\partial}{\partial y}\left(k\frac{\partial T}{\partial y}\right)+\frac{\partial}{\partial z}\left(k\frac{\partial T}{\partial z}\right)-$$

$$-\frac{\partial(uP)}{\partial x}-\frac{\partial(vP)}{\partial y}-\frac{\partial(wP)}{\partial z}+\frac{\partial(u\tau_{xx})}{\partial x}+\frac{\partial(u\tau_{yx})}{\partial y}+\frac{\partial(u\tau_{zx})}{\partial z}+\frac{\partial(v\tau_{xy})}{\partial x}+\frac{\partial(v\tau_{yy})}{\partial y}+\frac{\partial(v\tau_{zy})}{\partial z}+$$

$$+\frac{\partial(w\tau_{xz})}{\partial x}+\frac{\partial(w\tau_{yz})}{\partial y}+\frac{\partial(w\tau_{zz})}{\partial z}+\rho a.V \qquad \textbf{(3-5)}$$

Various fluid properties are required in these equations: ρ is density, V is velocity with its components in x, y, and z directions represented by u, v, and w (respectively), P is pressure, τ_{ij} is stress in the j direction exerted on a plane perpendicular to the i axis, b_i is i component of the body force per unit mass, U is internal energy, $\dot{q}$ is rate of volumetric heat addition per unit mass, k is thermal conductivity, and T is temperature.

The incompressible Navier-Stokes equations can be obtained from the compressible form simply by assuming that the density and viscosity are constant throughout the flow regime.

As explained by Churchill (1988), the momentum equations for a viscous flow were named the Navier-Stokes equations after Navier[5], who first derived these equations in 1822 on the basis of intermolecular arguments (Navier, 1822), and Stokes[6], who first derived these equations in 1845 for compressible fluid flow and without the molecular hypotheses of Navier (Stokes, 1845). The terminology of Navier-Stokes equations was then expanded to include the entire system of viscous flow equations (not only momentum equations).

---

[5] Claude-Louis Marie Henri Navier (1785-1836), a French civil engineer, was one of the first to develop a theory of elasticity.

[6] George Gabriel Stokes (1819-1903), an Irish mathematician and physicist, made extensive contributions to fluid mechanics

The Navier-Stokes equations are a coupled system of nonlinear partial differential equations, hence, are very difficult to solve. In fact, there is no general analytical solution to these equations. Finding effective approximation methods for solution of Navier-Stokes equations is at the heart of a broad range of engineering applications, from airplane design to nuclear reactor safety evaluation or to weather prediction (Elman et al., 2005).

As explained by Anderson (1995), because of some stability issues, the incompressible Navier-Stokes equations cannot be solved explicitly, and solution techniques for the incompressible equations are usually different from those used for solution of the Navier-Stokes equations for compressible flow. To overcome this difficulty, the pressure correction approach has been proposed. This accepted and widely used approach has been applied to both compressible and incompressible flows with good success (Anderson, 1995).

3.1.1.1 SIMPLE Method

The pressure correction approach is embodied in an algorithm called SIMPLE (Semi-Implicit Method for Pressure-Linked Equations). The SIMPLE algorithm was originally published in the often-cited paper by Patankar and Spalding (1972). The method was developed for incompressible fluid flow but has been extended successfully to compressible flows as well. The method has found widespread application over the past 30 years for both compressible and incompressible flows. The main features of the SIMPLE method are as follows:

1. The finite difference equations are obtained from the governing equations to form a coupled nonlinear system.
2. The pressure profile is estimated.
3. Using the values of pressure field, the momentum equations are solved for velocity profile.
4. The values of velocity profile will not necessarily satisfy the continuity equation. Hence, using the continuity equation the pressure profile is updated.
5. After obtaining the corrected pressure profile, return to step 3, and repeat the process until a velocity profile is found that does satisfy the continuity equation. When this is achieved, the Navier-Stokes equations have been solved.

The key point in this iterative method is obtaining a rational formula for correcting the pressure field. To obtain this, using pressure-velocity relationships from the linearized momentum equations, a pressure correction equation is derived from the continuity equation.

3.1.1.2 PISO Method

The Pressure-Implicit with Splitting of Operators (PISO) method is an improved version of SIMPLE algorithm. In the SIMPLE algorithm, updated velocities through the pressure correction equation do not necessarily satisfy the momentum equations. To improve the efficiency of the required iterative method, two complementary corrections, neighbor correction and skewness correction, are applied by the PISO method.

Issa (1986) noted that the main idea of the PISO algorithm is to bring the repeated calculations required by SIMPLE method inside the solution stage of the pressure correction equation. This iterative process is called "momentum correction" or "neighbor correction" and the updated velocities better satisfy the continuity and momentum equations, hence, rapidly move to convergence. Although the neighbor correction technique requires more CPU time at each iteration, it does reduce the number of iterations required for convergence, especially for dynamic simulation problems (Fluent 6.3 User's Guide, 2006). An internal iterative technique, similar to neighbor correction, is applied by PISO method for highly skewed meshes to reduce the impact of cell skewness on the quality of the updated velocities (Ferzieger and Peric, 1996). The technique, referred to as "skewness correction", significantly reduces convergence difficulties caused by highly skewed meshes.

The PISO method with neighbor correction has been recommended for all dynamic flow simulations, and this method with skewness correction has been recommended for highly skewed mesh systems (Fluent 6.3 User's Guide, 2006).

3.1.2 *Strategies for Simulation of Multiphase Fluid Flows*

There are two approaches to modeling multiphase flows: the Euler-Lagrange approach and the Euler-Euler approach. In the Euler-Lagrange approach, a continuous fluid phase is modeled by solving the time-averaged Navier-Stokes equations, and the dispersed phase is simulated by tracking a large number of particles through the flow field based on Newton's second law. The Euler-Euler approach, however, deals with the multiple phases as continuous phases that interact with each other. Since the volume of a phase cannot be occupied by the other phases, phase volume fractions are assumed to be continuous functions of space and time, and their sum is equal to one. Three different Euler-Euler multiphase models available in Fluent 6.3.26 are: the Volume of Fluid (VOF) model, the Mixture model, and the Eulerian model (Fluent 6.3 User's Guide, 2006).

3.1.2.1 Discrete Phase Model (DPM)

Following the Euler-Lagrange approach leads to the Discrete Phase Model (DPM) in Fluent. This model works well for flow regimes in which the discrete phase is a fairly low volume fraction, usually less than 12% (Fluent 6.3 User's Guide, 2006).

Various forces are taken into account while Fluent tracks the particles through the flow field that include the gravity force, the drag force (with the option of involving dynamic drag coefficient to account for particle deformation), the virtual mass force (accelerating the fluid surrounding the particle), the thermophoretic force (exerted on small particles suspended in a gas phase with a temperature gradient), the Brownian force, and the lift force.

Coalescence of particles and their breakups can also be modeled by DPM. For this purpose, based on the particle Weber number a proper model within the spray model theory is used.

3.1.2.2 VOF Model

The VOF model is a surface tracking model which is designed for simulation of immiscible multiphase flows where the position of the interface between any two adjacent different phases is of interest. In the VOF model, a single set of momentum equations is shared by the fluids, and the volume fraction of each of the fluids in each computational cell is tracked throughout the domain. Applications of the VOF model include free-surface flows, sloshing, the motion of large bubbles in a liquid, the motion of liquid after a dam break, the simulation of jet breakup, and the steady or dynamic tracking of any liquid-gas interface.

3.1.2.3 Mixture Model

In the Mixture model, the phases are assumed to be completely interpenetrating. The Mixture model solves for the mixture momentum equation and the dispersed phases are modeled via calculation of their relative velocities. Applications of the mixture model include bubbly flows, sedimentation, and cyclone separators.

3.1.2.4 Eulerian Model

The Eulerian model solves a set of momentum and continuity equations for each phase. The pressure and inter-phase exchange coefficients are defined among phases based on the type of phases involved. Applications of the Eulerian multiphase model include bubble columns, risers, and fluidized beds.

3.1.3 *CFD Modeling of Flow-Distributing Baffles and Wire Mesh Demisters*

The porous media model in Fluent 6.3.26 with appropriate modifications can be used to model the flow through baffles and demisters. This available model can also be used for a wide variety of equipment, such as modeling flow through packed beds, filter papers, perforated plates, flow distributors, and tube banks (Fluent 6.3 User's Guide, 2006). In order to use the model, a cell zone in which the porous media model is to be applied is defined and the pressure loss is modeled by setting appropriate parameters.

3.1.3.1 Momentum Equations for Porous Media

By using the porous media model, a momentum source equation is added to the governing momentum equations. The source term, Equation 3-6, is composed of two parts: a viscous loss term (Darcy), and an inertial loss term:

$$S_i = -\left(\sum_{j=1}^{3} D_{ij}\mu V_j + \sum_{j=1}^{3} C_{ij}\frac{1}{2}\rho|V|V_j\right) \qquad \textbf{(3-6)}$$

where S_i is the source term for the ith (x, y, or z direction) momentum equation, $|V|$ is the magnitude of the velocity and D and C are prescribed matrices. On the right hand side of Equation 3-6, the first term is the viscous loss term and the second term is the inertial loss term. The porous media momentum equation contributes to the pressure gradient in the porous cell, creating a pressure drop that is proportional to the fluid velocity or velocity squared in the cell.

In the case of a simple homogeneous porous media model, Equation 3-6 will reduce to Equation 3-7:

$$S_i = -\left(\frac{\mu}{\alpha}V_i + C_2\frac{1}{2}\rho|V|V_i\right) \qquad \textbf{(3-7)}$$

where α is the permeability factor and C_2 is the inertial resistance factor.

In laminar flow through porous media, the pressure drop is typically proportional to the velocity and the constant C_2 can then be set to zero. Ignoring convective acceleration and diffusion, the porous media model then reduces to Darcy's Law, Equation 3-8:

$$\nabla P = -\frac{\mu}{\alpha}\vec{V} \qquad \textbf{(3-8)}$$

At high flow velocities, the constant C_{ij} in Equation 3-6 provides a correction for inertial losses in the porous media. This constant can be viewed as a loss coefficient per unit length along the flow direction, thereby allowing the pressure drop to be specified as a function of dynamic head.

3.1.3.2 Modeling Baffles via the Porous Media Model

In modeling perforated plates, the permeability term in a porous media formulation can usually be eliminated and only the inertial loss term needs to be considered. This simplification leads to the following form of the porous media equation, Equation 3-9:

$$\nabla P = -\sum_{j=1}^{3} C_{2ij} \frac{1}{2} \rho |V| V_j \qquad \textbf{(3-9)}$$

Kolodzie and Van Winkle (1957), have shown that the following single-orifice based equation can be used as the flow equation through a perforated plate:

$$\dot{m} = C A_f \sqrt{\frac{2\rho \Delta P}{1 - \left(\frac{A_f}{A_p}\right)^2}} \qquad \textbf{(3-10)}$$

where, $\dot{m}$ is mass flow rate through the plate in *kg/s*, A_f is the total area of the holes in m^2, A_p is the total area of the plate in m^2, and C is the discharge coefficient.

Based on experiments, the discharge coefficient diagram has been provided as a function of Reynolds number, plate thickness, hole diameter, and hole pitch (Kolodzie and Van Winkle, 1957). Setting $\dot{m} = \rho V A_p$, Equation 3-10 can be rearranged to provide the pressure drop equation for a perforated plate, Equation 3-11:

$$\Delta P = \frac{1}{C^2} \left[\left(\frac{A_p}{A_f} \right)^2 - 1 \right] \left(\frac{1}{2} \rho V^2 \right) \qquad \textbf{(3-11)}$$

Dividing both sides of Equation 3-11 by the plate thickness, $\Delta x = \delta$, results in Equation 3-12:

$$\frac{\Delta P}{\Delta x} = \frac{1}{C^2 \delta} \left[\left(\frac{A_p}{A_f} \right)^2 - 1 \right] \left(\frac{1}{2} \rho V^2 \right) \qquad \textbf{(3-12)}$$

where V is the superficial velocity (not the velocity in the holes). Combining Equation 3-12 with Equation 3-9, results in Equation 3-13 for the constant C_2 in the direction normal to the plate:

$$C_2 = \frac{1}{C^2\delta}\left[\left(\frac{A_p}{A_f}\right)^2 - 1\right] \qquad \textbf{(3-13)}$$

Therefore, if the configuration and dimensions of flow-distributing baffles are provided, Equation 3-13 can be used to calculate C_2, and the porosity of baffles, which is necessary for the porous media model, can also be calculated.

3.1.3.3 Modeling Wire Mesh Demisters via the Porous Media Model

Knitted wire mesh demisters can be modeled as a porous media. For this purpose, the porous media parameters, which are used for pressure drop calculations in the media, need to be set. Since the mesh pad demisters generally result in very low pressure drops, their pressure drop was not modeled and assumed to be negligible. Fortunately, just recently, a comprehensive and practical research study has appeared in the public literature that deals with the characterization of pressure drop in knitted wire mesh demisters. Helsør and Svendsen (2007) have reviewed the two other relevant studies in this field and presented their model for pressure drop calculation in mesh pads. In their experimental studies, the data have been collected and analyzed for seven different wire mesh demisters, at four different system pressures (ranging from atmospheric pressure to 9.2 *MPa*), and with three different fluids (air, nitrogen, and natural gas). The data have been fit to a Hazen-Dupuit-Darcy type equation, Equation 3-14:

$$\frac{\Delta P}{\Delta x} = \frac{\mu}{\alpha}V + C\rho V^2 \qquad \textbf{(3-14)}$$

and the correlating parameters (α and C) have been provided for the different mesh pad types. Comparing Equation 3-14 with the momentum source equation for porous media and its simplified form (Equations 3-6 and 3-7), results in Equation 3-15 for the constants D and C_2 in a flow direction normal to the mesh pad:

$$D = \frac{1}{\alpha} \; ; \; C_2 = 2C \qquad \textbf{(3-15)}$$

Therefore, provided that the type and characteristics of mesh pad are available, Equation 3-15 can be used through the correlations developed by Helsør and Svendsen (2007) to calculate the parameters required for the porous media model.

3.2 Simulation of Pilot-Plant-Scale Two-Phase Separators

In this section, the steps and input data required for CFD simulation of pilot-plant-scale two-phase separators are presented. The paper titled "Analytical Study of Liquid/Vapor Separation Efficiency", by Monnery and Svrcek (2000), and a field pilot plant skid at the Prime West East Crossfield gas plant have been used as the basis for the CFD model.

3.2.1 *Experimental Equipment and Procedure*

The process schematic of the experimental equipment is shown in Figure 3-1. The system consists of two-phase separators, gas inlet piping, liquid pumping and injection, and a high efficiency filter to collect entrained separator liquids. The aim of experiments was studying phase separation performance in the three horizontal separators and one vertical separator operating at three different pressures 70 kPa, 700 kPa, and 2760 kPa.

The experimental procedure consisted of setting the gas pressure using the outlet manual globe valve and the flow was adjusted using the inlet globe valve to have established the desired flow-rate at the desired pressure. Once the gas flow stabilized, the liquid phase was injected in an amount that over-saturated the gas. The system was left until the flows and temperatures showed a steady-state operating condition.

The experimental data consisted of separator liquid level, gas flow-rate, and liquid carryover. These data were then used to determine the gas velocity and corresponding separation efficiency. The incipient velocity of gas phase which results in liquid droplet carryover was measured and the value was used to estimate the entrained droplet diameter.

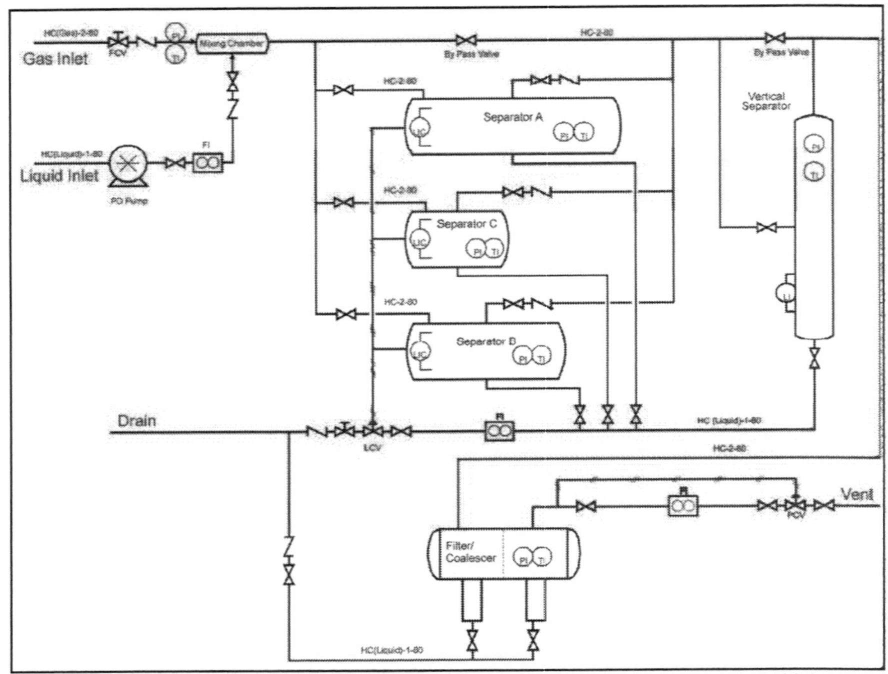

Figure 3-1. Process Flow Diagram for Separator Skid (Monnery and Svrcek, 2000).

3.2.2 *Material Definition*

The composition of saturated gas at three different pressures, as provided by Monnery and Svrcek (2000), is given in Table 3-1. To determine the saturated liquid composition and other physical properties, the Peng-Robinson (PR) equation of state was used in the commercial process simulator, HYSYS 3.2. The predicted values used in the CFD simulations are presented in Table 3-2.

Table 3-1. The composition of the saturated gas at multiple pressures (Monnery and Svrcek, 2000).

Component	Mole Fraction		
	70 kPa	700 kPa	2760 kPa
Nitrogen	0.0377	0.0446	0.0454
Methane	0.7573	0.8958	0.9113
Ethane	0.0183	0.0217	0.0221
Propane	0.0055	0.0065	0.0066
i-Butane	0.0231	0.0041	0.0020
n-Butane	0.0766	0.0132	0.0060
i-Pentane	0.0298	0.0051	0.0022
n-Pentane	0.0271	0.0046	0.0021
n-Hexane	0.0157	0.0028	0.0014
n-Heptane	0.0029	0.0005	0.0003
n-Octane	0.0005	0.0001	0.0001
n-Nonane	0.0001	0.0000	0.0000
Methylcyclopentane	0.0018	0.0003	0.0001
Cyclohexane	0.0014	0.0003	0.0001
Methylcyclohexane	0.0008	0.0001	0.0001
Benzene	0.0009	0.0002	0.0001
Toluene	0.0005	0.0001	0.0001
p-Xylene	0.0001	0.0000	0.0000

Table 3-2. The values of material properties used in simulation case studies.

		Case Study (I)	Case Study (II)	Case Study (III)
Operating Pressure (kPa)		70	700	2760
Continuous Phase	Density (kg/m^3)	1.8992	6.6164	23.846
	Viscosity ($Pa.s$)	1.062×10^{-5}	1.0812×10^{-5}	1.1473×10^{-5}
Discrete Phase	Density (kg/m^3)	691.52	690.04	670.05
	Viscosity ($Pa.s$)	4.0934×10^{-4}	4.0882×10^{-4}	3.3683×10^{-4}
	Surface Tension (N/m)	0.019905	0.019676	0.01707

3.2.3 *Implementing DPM Approach*

The first approach used to simulate vapor-liquid separation in the two-phase separators was the DPM multiphase model available in Fluent 6.3.26. In order for this model to provide useful results, the dispersed phase should be a very low volume fraction, less than 12% (Fluent User's Guide, 2006). This was the situation that existed in the vapor phase of the two-phase separators in question. The overall strategy adopted was that of Newton et al. (2007), in which only separation of liquid dispersion from vapor phase was simulated. It was also assumed that the gas-liquid interface in the separators was defined as a frictionless wall which traps the droplets coming into contact with it.

3.2.3.1 Physical Models

The physical models consist of four separators labeled vessel A, B, C (all horizontal) and D (vertical). Figure 3-2 provides the individual vessel specifications. Mesh generation step was completed using Gambit 2.4.6, the Fluent preprocessor tool. In order to have a good grid quality, the external edges of nozzles were discretized at first. Then the faces and the whole volume were swept by the Gambit mesh generation tool. The supplementary package of the book includes screen photos from the Gambit environment during the development of the grid systems for all the separators. However, for convenience, Figure 3-3 and Figure 3-4 provide example photos for separators B and D, respectively.

Quality of the produced mesh was examined based on the cell skewness factor. The results, outlined in Table 3-3, indicate that a negligible fraction of cells, less than 0.01%, is of poor quality. However, the grids with a cell skewness factor above 0.8 were converted to polyhedral grids by employing a new capability of the latest version of Fluent. The values of maximum cell squish, maximum skewness, and maximum aspect ratio for original and modified mesh are reported in Table 3-4.

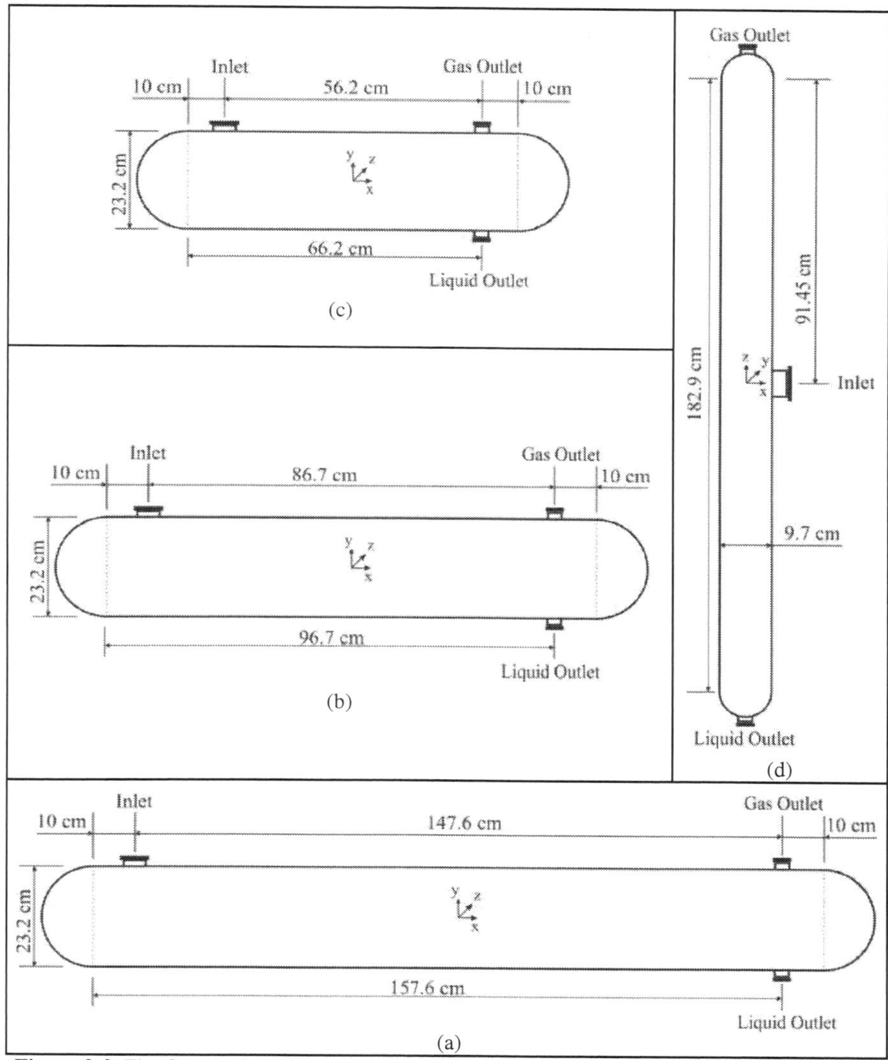

Figure 3-2. The Geometrical Specifications of Models; (a) Vessel A, (b) Vessel B, (c) Vessel C, (d) Vessel D.

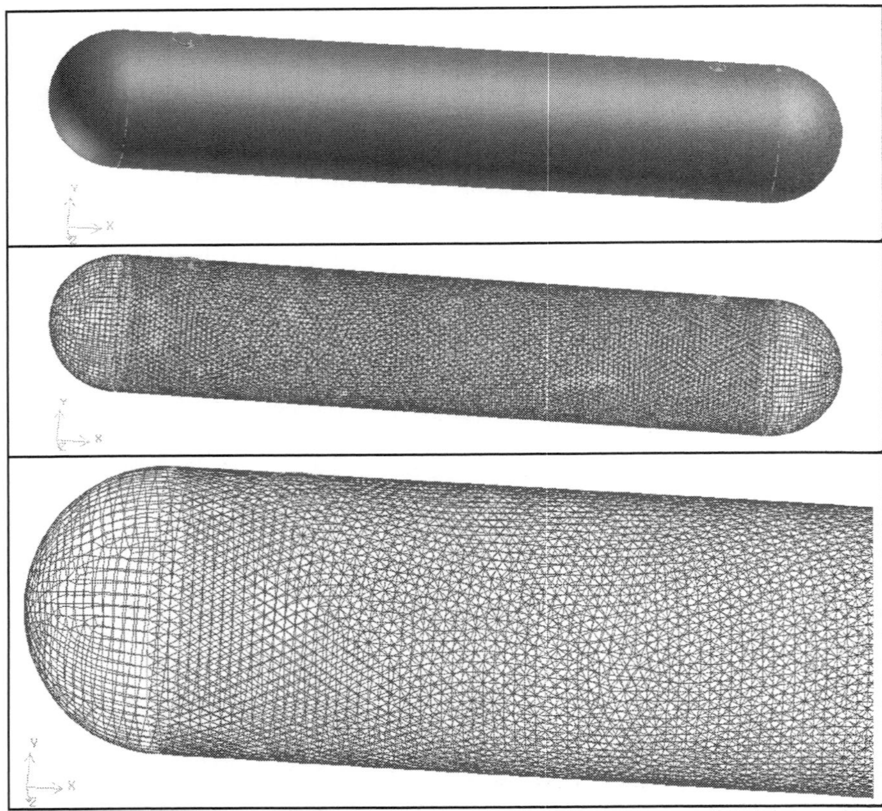

Figure 3-3. The Grid System for Vessel B Produced in Gambit Environment (DPM Approach).

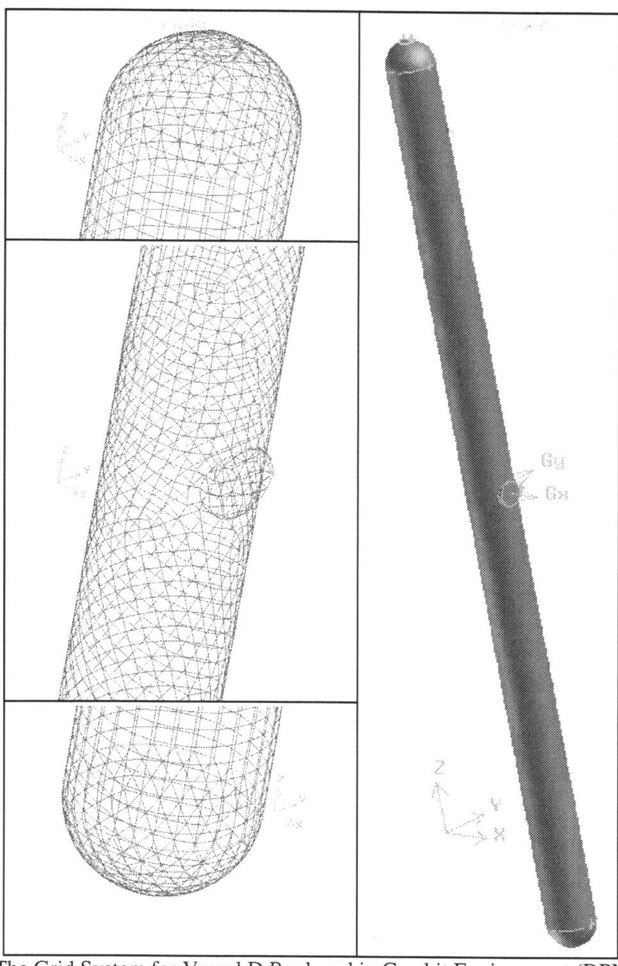

Figure 3-4. The Grid System for Vessel D Produced in Gambit Environment (DPM Approach).

Table 3-3. Quality of the mesh produced in Gambit environment for two-phase separators (DPM approach).

	Skewness				
Model	0-0.20	0.20-0.40	0.40-0.60	0.60-0.80	0.80-1.0
A	29.56%	47.22%	19.49%	3.72%	0.00%
B	29.56%	46.47%	19.96%	4.01%	0.00%
C	30.05%	45.64%	19.79%	4.51%	0.01%
D	39.30%	34.42%	20.64%	5.64%	0.01%

Table 3-4. Global quality of the original and modified mesh for two-phase separators (DPM approach).

		Number of Cells	Maximum Squish	Maximum Skewness	Maximum Aspect Ratio
A	Original	320602	0.787114	0.829467	22.1483
	Converted	320597	0.779131	0.829057	22.1483
B	Original	204008	0.787114	0.829467	22.1484
	Converted	204003	0.779131	0.829057	22.1484
C	Original	143064	0.770505	0.821138	20.1374
	Converted	143050	0.770505	0.820576	19.8758
D	Original	62903	0.770509	0.821137	20.1372
	Modified	62900	0. 770509	0.805442	19.0316

3.2.3.2 Definition of Droplet Size Distribution

As noted, the experimental data contained the incipient carryover velocity of gas phase for multiple separator operating pressures. These values were then used to estimate the entrained liquid droplet size. In this CFD simulation study, however, the probable droplet size distribution was input and the incipient gas velocity that resulted in the liquid carryover was determined. The common particle size distribution function of Rosin-Rammler (1933), Equation 3-16, was used:

$$Y_d = \exp\left[-\left(\frac{d}{\bar{d}}\right)^n\right] \qquad \textbf{(3-16)}$$

where Y_d is the mass (or volume) fraction of droplets with diameter greater than d. The Rosin-Rammler equation contains two parameters: volume mean diameter, $\bar{d}$, and spread parameter, n, which specifies how narrow the distribution around the mean diameter is.

Monnery and Svrcek (2000) presented maximum stable and average entrained liquid droplet diameter at two different pressures 70 and 700 kPa. Based on their results and assuming a minimum droplet size of 100 μm (which is small enough for simulation purposes), the discrete phase parameters were set to the values shown in Table 3-5. As Table 3-5 indicates, three different values were assumed as the mean droplet size to study the effect of this parameter on the predicted incipient velocity. These values were 150 μm (the well-known design droplet size), 550 μm (the average of experimental mean droplet sizes), and 1000 μm.

3.2.3.3 Setting CFD Simulator Parameters

After importing the mesh file and making the modifications, the necessary material properties for continuous and discrete phase were input. The discrete phase model could then be specified by setting the droplet size distribution, the surface from which droplets were injected to the vessel (nozzle surface), and the velocity of injection which was assumed to be equal to the continuous phase velocity.

In all case studies the Reynolds number was much more than the transient value (Re = 2300). Therefore, a suitable turbulence model was selected as the viscous model. In this study, the standard $k-\varepsilon$ (Launder and Spalding, 1972) model was selected because of its robustness, economy and accuracy for a wide range of turbulent flows in industrial flow simulations (Fluent 6.3 User's Guide, 2006). This semi-empirical model has been accepted as the most cost-effective and widely applicable turbulence model and has been selected as the default in most commercial packages (Sharratt, 1990; Gosman, 1998).

Proper boundary condition setting was very important in this study. For the continuous phase, inlet velocity and outlet pressure were set. The values of inlet velocity were varied in order to find the incipient velocity. When setting the flow regime in the inlet and outlet nozzles, the turbulence intensity and hydraulic diameter of the nozzles were determined. For calculation of the turbulence intensity, the empirical correlation for pipe flows, as recommended by Fluent 6.3 guidelines, was used:

$$I = 0.16\,\mathrm{Re}^{-0.125} \qquad \textbf{(3-17)}$$

In the developed CFD model, the droplets reaching the gas-liquid interface were assumed to become part of the liquid phase. In fact, this assumption tends to simplify the separation task for liquid droplets and might lead to higher separation efficiencies. In order to compensate partially, the interface levels were specified somewhat lower than in normal design practice. Therefore, the interface level in the vessels was assumed to be around 5 *cm* from the vessel bottom. When a droplet contacted with walls other than defined interface, it was assumed that 95% of its normal momentum and 90% of its tangential momentum were lost. Thus, normal and tangent reflection coefficients were set to be 0.05 and 0.10, respectively.

Careful choice of the solution method and under-relaxation factors is a major contribution toward both the rate of convergence and the solution existence (Sharratt, 1990; Anderson, 1995). Following the Fluent 6.3 guidelines and the trend of solution convergence, the solver parameters were set as follows:

Discretization Method for Momentum: Second Order Upwind
Solution Method: SIMPLE
Under-Relaxation Factor for Pressure ≈ 0.9
Under-Relaxation Factor for Momentum ≈ 0.1

Table 3-5. The discrete phase parameters used for simulation of two-phase separators through DPM approach.

Number of Tracked Particles	d_{min} (μm)	d_{max} (μm)	$\overline{d}$ (μm)	n	Total Flow-rate (kg/s)
1000	100	3000	150 550 1000	1.0	6×10^{-5}

3.2.4 *Implementing VOF-DPM Approach*

As the second approach to simulating vapor-liquid separation in the two-phase separators, an effective combination of DPM and VOF multiphase models within Fluent 6.3.26 was used. This approach was selected considering the nature of the empirically studied phase separation process and the features of multiphase models available in Fluent. Consequently, the VOF model was used to obtain the overall picture of the fluid flow behavior in the two-phase separators, and then, in order to move the simulation toward a realistic situation, liquid droplets were injected from the inlet to be tracked by DPM model. The equations governing the injected discrete phase (liquid droplets) and coexisting continuous phases (gas and liquid phases) were solved simultaneously. This approach required significant computational time because after having the continuous fluid phases converged to their preliminary solutions, interactions among multiple discrete and continuous phases were included.

3.2.4.1 Grid Systems

In Gambit environment, the external edges of nozzles were first discretized, and the faces and all the volume were then swept. The screen photos from the Gambit environment while developing the grid systems for all the separators are provided in the supplementary package of the book, and for convenience, Figure 3-5 and Figure 3-6 include such photos for separators C and D, respectively. The quality of the produced mesh was examined based on the cell skewness factor. The results, Table 3-6, show that a negligible fraction of cells (less than 0.01%) is of poor quality. However, in order to reach an even better grid quality, the grids with a cell skewness factor above 0.8 were converted to polyhedral grids in the Fluent environment. The values of maximum cell squish, maximum skewness, and maximum aspect ratio for original and modified mesh are reported in Table 3-7.

3.2.4.2 Definition of Droplet Size Distribution

Droplet size distribution was defined similar to the DPM approach, presented in the previous section (Table 3-5), with the difference for this phase being the mean droplet size of 150 μm.

This size is an industrial accepted separator design droplet size, and its suitability in prediction of incipient velocities was also confirmed using the DPM approach[7].

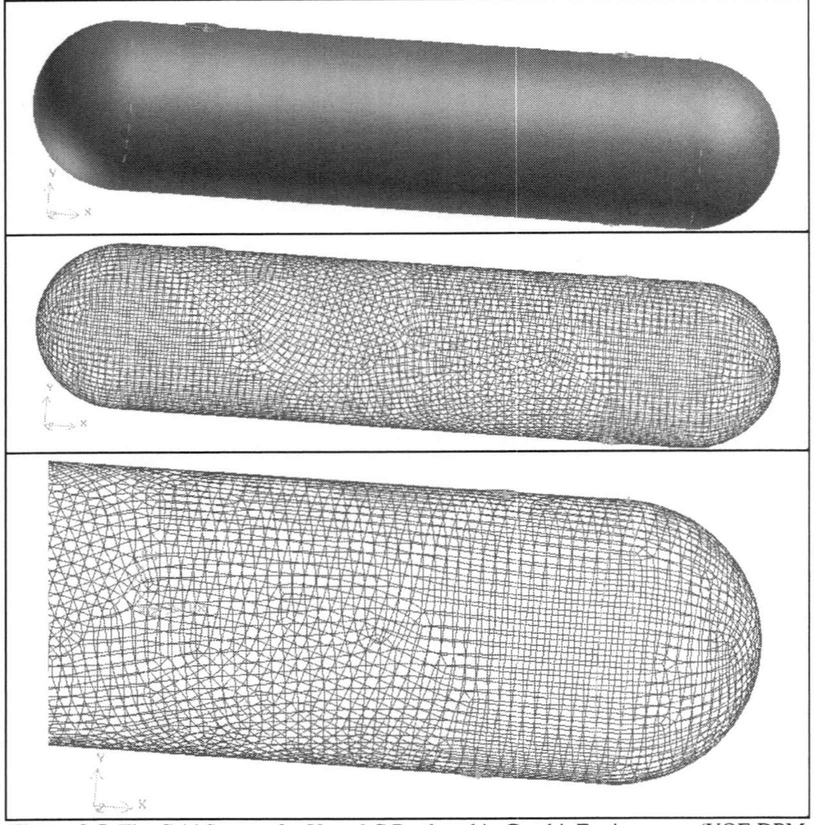

Figure 3-5. The Grid System for Vessel C Produced in Gambit Environment (VOF-DPM Approach).

[7] Please refer to Chapter Four for details.

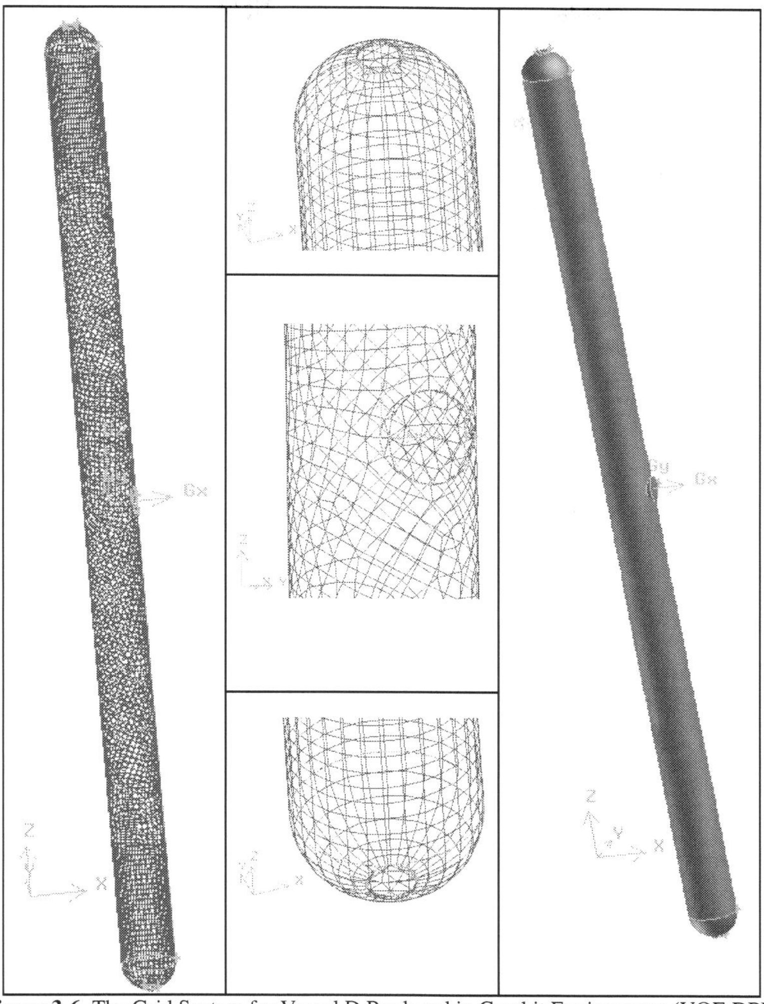

Figure 3-6. The Grid System for Vessel D Produced in Gambit Environment (VOF-DPM Approach).

Table 3-6. Quality of the mesh produced in Gambit environment for two-phase separators (VOF-DPM approach).

	Skewness				
Model	0-0.20	0.20-0.40	0.40-0.60	0.60-0.80	0.80-1.0
A	29.58%	46.72%	19.81%	3.89%	0.004%
B	29.97%	46.34%	19.75%	3.93%	0.004%
C	29.66%	46.12%	20.10%	4.11%	0.01%
D	38.30%	36.27%	20.75%	4.60%	0.08%

Table 3-7. Global quality of the original and modified mesh for two-phase separators (VOF-DPM approach).

		Number of Cells	Maximum Squish	Maximum Skewness	Maximum Aspect Ratio
A	Original	326550	0.790256	0.829117	22.9059
	Converted	326530	0.776824	0.820576	19.8753
B	Original	219586	0.787082	0.829389	22.1426
	Converted	219579	0.778983	0.828974	22.1426
C	Original	165921	0.777096	0.821001	20.1331
	Converted	165914	0.776824	0.820576	19.8753
D	Original	68174	0.770605	0.888568	30.0719
	Modified	68114	0.770605	0.884907	19.0413

3.2.4.3 Setting CFD Simulator Parameters

The CFD simulator parameters were set after importing the mesh file into Fluent 6.3.26 environment. Similar to DPM approach, the material properties were input and discrete phase was specified by setting its droplet size distribution and injection characteristics. Again, the standard $k-\varepsilon$ model was selected for turbulence modeling.

The liquid depth in the separators was set using the normal practice as described by Smith (1987)[8]. Therefore, the depth of liquid in vertical separator was set to be 29.1 cm, and in horizontal separators was set to be 6.6 cm. In order to initiate the iterative solution procedure, the position of interface between the gas and liquid phases was specified. For this purpose, the volume fractions of phases above and below the assumed interface plane were set using the

[8] Please refer to "Overall Aspects of Separator Geometry" in "Two-Phase Separators" subsection of Chapter Two for more details.

"Patching" tool of Fluent. Since the solver might distort the shape and position of the interface surface, the iterations were stopped at frequent intervals and the interface level was checked and corrected (if necessary) by patching the volume fractions of phases. The liquid level must be maintained at the correct height during the iterative solution process.

At the boundary setting stage, the inlet velocity and volume fractions were specified for inlet continuous phases. Then, the outlet pressure and volume fractions (as pure gas) were set for the gas-outlet boundary, and the outlet velocity and volume fractions (as pure liquid) were set for the liquid-outlet boundary.

In the experimental research project, the amount of liquid flow-rate was constant while the gas flow-rate was changed to provide different superficial velocities in the separators (Monnery and Svrcek, 2000). The flow-rate of liquid was assumed to be 3.59×10^{-4} m^3/s which was corresponding to occupying 1% of inlet nozzle surface area by liquid phase to give an overall 0.10 m/s superficial velocity in horizontal separators. Hence, the liquid outlet velocity was set to 0.7811 m/s. Note, the values of inlet velocities and volume fractions were always set so that the liquid flow-rate was maintained constant. For example, in order to have a superficial gas velocity of 0.5 m/s in the horizontal separators, the mixture inlet velocity and the liquid volume fraction were set to 9.11 m/s and 0.0198. For the same superficial velocity in the vertical separator, the mixture inlet velocity and the liquid volume fraction were set to 2.02 m/s and 0.089. Similar to the DPM approach, the turbulence intensity in the inlet and outlet zones was estimated using Equation 3-17.

For discrete phase modeling, the destiny of droplets after contacting with internal walls of vessels must also be specified. For this purpose, it was assumed that the droplets reaching the bottom of vessels (containing liquid) were drained, that is not returning to the vapor phase. Furthermore, it was assumed that 95% of the normal momentum and 90% of tangential momentum of a droplet contacting with vessel walls (other than bottom wall) was lost. Thus, normal and tangent reflection coefficients were set to 0.05 and 0.10, respectively.

The process of solving a multiphase simulation problem is inherently difficult, and stability and/or convergence problems would be encountered (Fluent User's Guide, 2006). The Fluent 6.3 guidelines and the solution convergence trend were used to set the solver parameters so as to overcome stability and convergence problems:

Discretization Method for Pressure: Body Force Weighted Discretization Method for Momentum: First Order Upwind Solution Method: PISO Under-Relaxation Factor for Pressure = 0.1 Under-Relaxation Factor for Momentum = 0.001 Under-Relaxation Factor for Volume Fraction = 0.005 Under-Relaxation Factor for Turbulent Groups (Kinetic Energy, Dissipation Rate, and Viscosity) = 0.7

3.3 Simulation of a Large-Scale Three-Phase Separator

A three-phase separator that is part of a large North Sea production platform was next simulated. The separator is located in the Gullfaks oil field in the Norwegian sector of the North Sea. The production on the Gullfaks-A platform started in 1986-87. Figure 3-7 presents the details, as provided by Hanson et al. (1993), of the separator as it was designed and installed on the platform. Figure 3-8, taken from Hansen et al. (1993), shows the flow diagram for the production process. Represented in Figure 3-8, the separator of interest is the first stage of the three-stage dual-train production process.

The separator operates at a fixed pressure of 6870 *kPa*. The inlet multiphase fluid flow enters the vessel as a high intensity momentum jet that strikes the inlet momentum breaker (deflector baffle). The liquid drops into the liquid pool in lower part of the vessel and the gas together with liquid droplets flows in the upper part of the vessel. In the vapor phase, liquid droplets will settle to the liquid interface, flowing to the corresponding liquid phase outlets. The upper part of the vessel is equipped with internals, including distribution baffles and demister, to enhance the separation of liquid droplets from vapor.

During the first four years of operation, separation inefficiencies have been experienced due to the following operating problems:

- water level control failure with increasing production of water,
- emulsion problems inside separator,
- sand accumulation, and
- increased water content in the oil outlet.

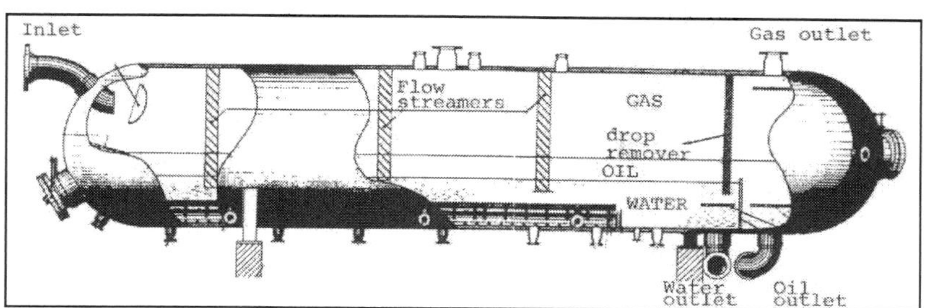

Figure 3-7. General Configuration of the First Stage Separator on the Gullfaks-A Production Facility (Hansen et al., 1993).

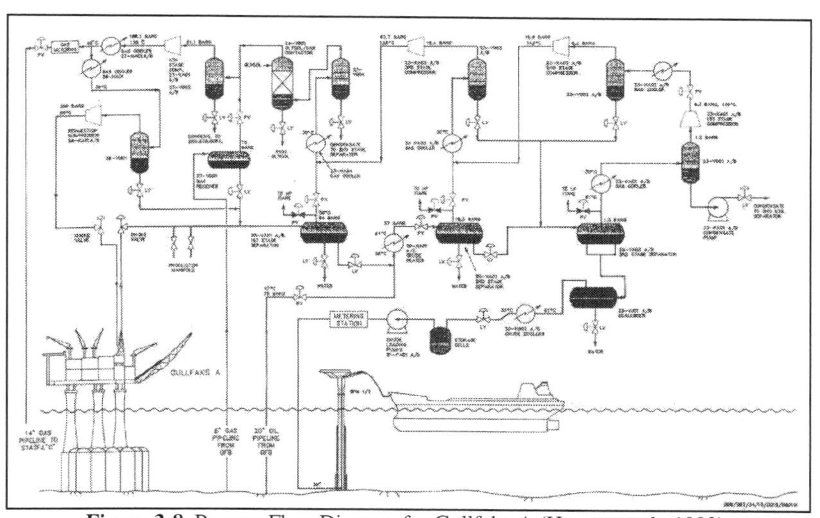

Figure 3-8. Process Flow Diagram for Gullfaks-A (Hansen et al., 1993).

In order to develop a deep understanding of the complex three-phase separation process Hansen et al. (1993) decided to model the overall fluid flow regimes inside the separator. To simplify this complicated simulation task, they focused on two zones: the inlet and momentum breaker zone and the bulk liquid flow zone. Based on operational experience and the simulation results, Hansen et al. (1993) proposed modifications to improve the separation efficiency that included redesigning the sand removal system, modifying the liquid level measurement device, and reducing the height of the demister and baffles to be completely out of the liquid phase. The modifications to the separator did improve both the water level control and the produced oil quality. In the current CFD simulation study, all the separation zones of the separator will be simulated. Therefore, the results should provide an overall picture of separation quality not only in the inlet and bulk liquid zones but also in the gas and interface zones. Exploiting the various multiphase models available in Fluent, the combined VOF-DPM model is the best choice for modeling both the macroscopic and microscopic features of the three-phase separator and was used to model the phase separation behavior in this field separator. The necessary model settings will be described in detail in the following sections.

3.3.1 *Material Definition*

The physical parameters for the fluids in the Gullfaks-A separator are taken from Hansen et al. (1993) and presented in Table 3-8. In the CFD simulation, the vessel was assumed to operate at a constant pressure, hence the pressure drops of the fluid phases flowing through the baffles and the demister pad were assumed to be negligible. Note, in Chapter Four, this criterion will be utilized to physically verify the modeling of the baffles and the mist eliminator.

The other important physical properties necessary for CFD simulation of the three-phase separation are interfacial surface tensions. Since these data were not given in the original paper, estimated values were used in the current study. For this purpose, a four component mixture of n-$C_{27}H_{56}$, n-$C_{28}H_{58}$, n-$C_{29}H_{60}$, and n-$C_{30}H_{62}$, with mole fractions of 0.85, 0.05, 0.05, and 0.05, respectively, were used in HYSYS 3.2 to simulate the oil phase. The criterion for setting the composition of the mixture was the accuracy of the mixture density and viscosity at operating temperature and pressure compared to the values given in the original study. Using the PR equation of state and the TRAPP model, density and viscosity of the mixture were estimated to be 783.59 kg/m^3, and 0.005296 *Pa.s*, respectively. The estimation errors for the oil density and viscosity were 5.76% and 0.88%, respectively. Thus, it was assumed that the oil-gas surface tension estimated by HYSYS should be reasonable, and a surface tension of 0.0238 *N/m* was assumed for the oil-gas interface. The assumed value compared well with the oil surface tension range of 0.023 to 0.038 *N/m* at 20°*C* proposed by Streeter and Wylie (1985). HYSYS was also used for estimation of water-gas surface tension, and the estimated value was 0.0668 *N/m*. Using the chart provided by Heidemann et al. (1987), surface tension of pure water is 0.067 *N/m* at 55.4°*C* which is in agreement with the HYSYS estimate. Finally, an empirical study was used to estimate oil-water surface tension: Kim and Burgess (2001) have reported a value of 0.052 *N/m* as the surface tension for a mineral oil and water interface at 25°*C*. They noted that although oil is a mixture of various hydrocarbons, each constituting hydrocarbon in contact with water, has almost the same interfacial surface tension. Thus, it was assumed that the surface tension for oil-water interface at 25°*C* would be almost the same as reported by Kim and Burgess (2001). Furthermore, in order to account for surface tension temperature functionality, the reported value was modified using the method proposed by Poling et al. (2001), Equation 3-18, which provides a correlation between the surface tension and the reduced temperature, T_r:

$$\frac{\sigma(T_{r2})}{\sigma(T_{r1})} = \left(\frac{1-T_{r2}}{1-T_{r1}}\right)^{1.2} \qquad \textbf{(3-18)}$$

Note, while using Equation 3-18 for alcohols, the right hand side exponent should be changed to 0.8. Using Equation 3-18 and assuming a pseudo-critical temperature of 576°*C* for the oil as estimated by HYSYS, a surface tension of 0.0486 *N/m* was estimated for the oil-water interface at 55.4°*C*. The estimated value is in agreement with the Antonoff's rule in that the oil-water surface tension should approximately be equal to the absolute difference between oil and water surface tensions (Antonoff, 1907), which are 0.0238 *N/m* and 0.0668 *N/m*, respectively.

Table 3-8. The physical parameters of fluids in Gullfaks-A separator provided by Hansen et al. (1993).

	1988 Production Rate (m^3/h)	Future Production Rate (m^3/h)	Density (kg/m^3)	Viscosity ($Pa.s$)
Gas	1640	1640	49.7	1.30×10^{-5}
Oil	1840	1381	831.5	5.25×10^{-3}
Water	287	1244	1030	4.30×10^{-4}
Operating Conditions	Temperature = 55.4°*C* and Pressure = 6870 *kPa*			

3.3.2 *Physical Model*

Figure 3-9 provides the specifications of Gullfaks-A separator and does show that not all the required dimensions were provided by Hansen et al. (1993) in their paper. The missing information, however, may be estimated using the given dimensions for the realistic schematic of the separator shown in Figure 3-7. To complete the information required for a CFD simulation, the inlet nozzle and the outlet nozzles were assumed to have a diameter of 0.48 *m* and 0.38 *m*, respectively, while the diameter of hemispherical momentum breaker and the height of weir were assumed to be 0.80 *m* and 1.25 *m*, respectively. Generation of the corresponding mesh system was performed in the Gambit 2.4.6 environment. Screen photos of the generated grid system in the Gambit environment have been included in Figure 3-10. In order to have a discretized model with "good" grid quality, the mesh generation process should be completed in a step by step sequence. The vessel was split into areas and the inlet nozzle, momentum breaker, splash plate, weir, and outlet nozzles were first discretized. In doing so, the edges of nozzles and other internals were discretized before the mesh generation for the separator surfaces and volumes. Then, the cylindrical part of the vessel was discretized such that some cells in this part were separated and referred to as the porous media and did include mesh for distribution baffles and the demister pad. The horizontal surrounding surfaces of each baffle (with thickness of 0.02 *m*) and those of demister pad (with thickness of 0.15 *m*) were assumed to be flat surfaces. Therefore, in the cylindrical part of vessel, the grids must be fine enough and arranged horizontally in regular and constant intervals. After generating the mesh for the cylindrical part of the vessel, the remaining parts of the vessel were "swept" by the Gambit mesh generation tool.

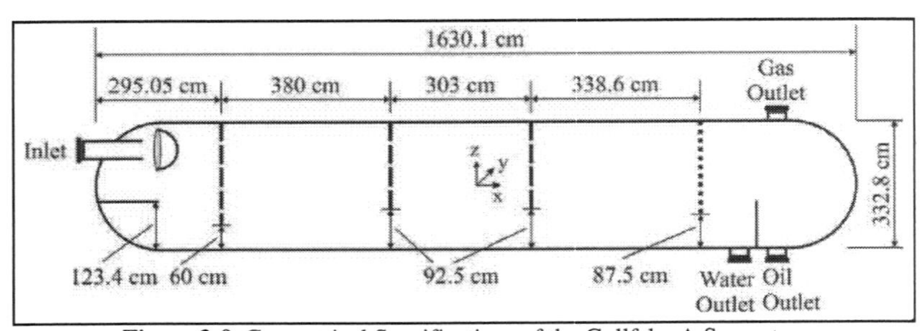

Figure 3-9. Geometrical Specifications of the Gullfaks-A Separator.

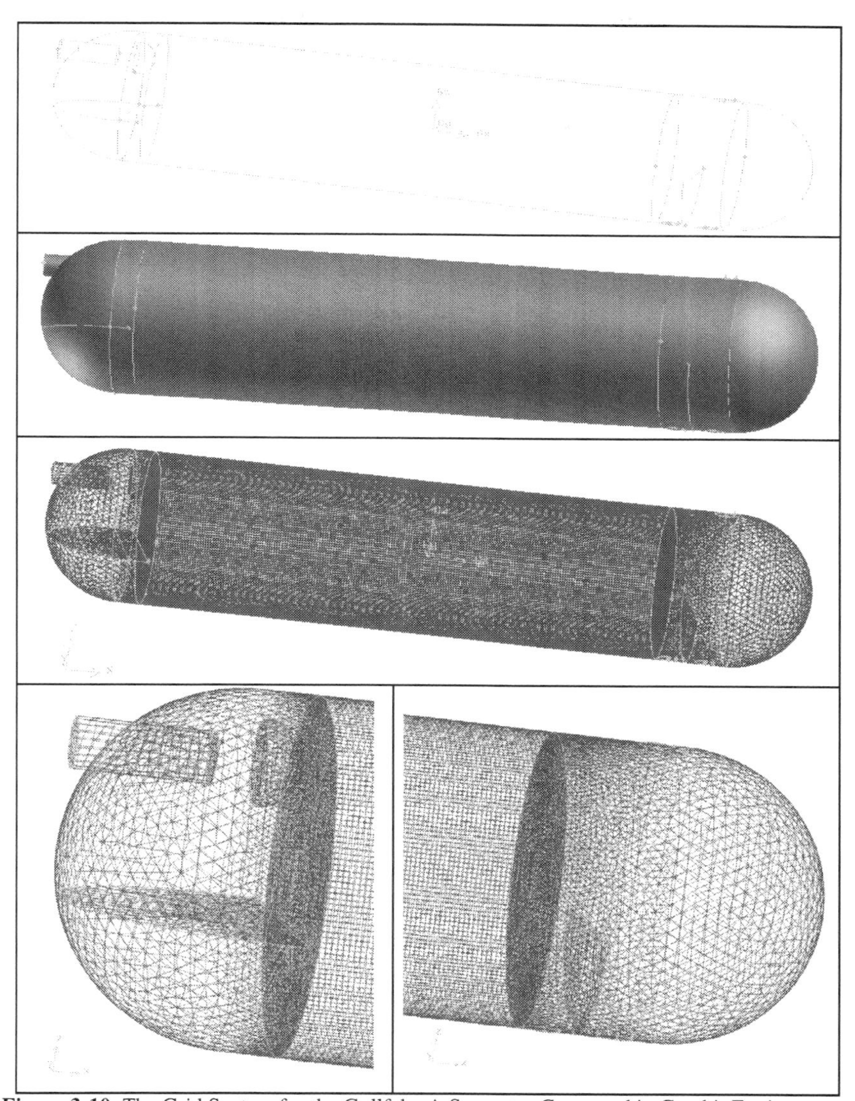

Figure 3-10. The Grid System for the Gullfaks-A Separator Generated in Gambit Environment.

Table 3-9. Quality of the mesh produced for the Gullfaks-A separator in Gambit environment.

Number of Cells	Maximum Squish		Maximum Skewness		Maximum Aspect Ratio
884847	0.748182		0.895873		44.1708
Skewness of the produced mesh					
Skewness Range	0-0.20	0.20-0.40	0.40-0.60	0.60-0.80	0.80-1.0
Density of Cells	79.0416%	15.4785%	3.8489%	1.6285%	0.0025%

The global quality of the produced mesh in terms of number of cells, maximum cell squish, maximum skewness, and maximum aspect ratio are presented in Table 3-9. Furthermore, to ascertain the quality of the generated mesh, the cell skewness was evaluated. As shown by the mesh results of Table 3-9 only a negligible fraction of cells (0.0025%) was of poor quality. However, by employing a new feature of Fluent 6.3.26, the grids with cell skewness factor above 0.8 were converted to polyhedral grids. Although this minor modification did not reduce the maximum values reported in Table 3-9, the number of cells was reduced from 884847 to 884805.

3.3.3 *Modeling the Baffles and the Mesh Pad Demister*

As described in section 3.1.3, the porous media model can be used to model the flow through baffles and the demister. This section will explain how the required adjustments were made in the Fluent 6.3.26.

3.3.3.1 Modeling the Baffles

As noted in section 3.1.3.2, the configuration and dimensions of flow-distributing baffles are taken into account for calculation of C_2 and ε parameters which are necessary for the porous media model. However, these crucial specifications were not provided in the original paper of Hansen et al. (1993). Fortunately, Hansen et al. (1991) have presented useful experimental data previously. Hansen et al. (1991) performed several experiments to obtain data in order to validate the results of their developed computer code, FLOSS, which was used for the simulation of Gullfaks-A separator. Inside their experimental model, with dimensions of 0.46 m×0.46 m×1.83 m, the flow-distributing baffle was specified as a perforated plate with 173 holes, each with diameter of 6.4 mm and distance between centers of 25 mm. Since the model has been used for validation of the computer code, and the computer code was used to simulate the Gullfaks-A separator, it can be expected that the model baffle had the same overall configuration (hole pattern) as the baffles of Gullfaks-A separator. Therefore, assuming a thickness of 20 *mm* for distribution baffles, the porosity (ε), and constant C_2 could be calculated using Equations 3-19 to 3-23:

Calculation of ε

From Figure 3-11, we have $$\varepsilon = \frac{A_f}{A_p} = \frac{\frac{\pi d^2}{4}}{a^2} = \frac{\frac{\pi (6.4)^2}{4}}{25^2} \approx 0.05 \qquad \textbf{(3-19)}$$

Calculation of C_2

First, *HolePitch* was calculated based on Figure 3-11:

$$a = 25\ mm \Rightarrow HolePitch = 25\sqrt{2} = 35.355\ mm \qquad \textbf{(3-20)}$$

Then Re number was calculated for gas flow through the baffle holes:

$$\mathrm{Re} = \frac{Vd}{\nu} = \frac{2.095 \times 0.0064}{2.6157 \times 10^{-7}} = 51270\ (> 4000) \qquad \textbf{(3-21)}$$

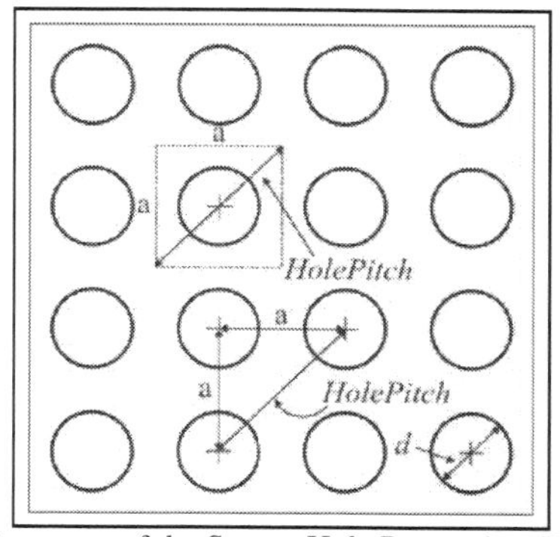

Figure 3-11. Geometry of the Square Hole Pattern in a Perforated Plate.

Considering that $\frac{\delta}{d} = \frac{20}{6.4} = 3.125$, and using the correlation developed by Kolodzie and Van Winkle (1957), C could be calculated as:

$$C = 0.98\left(\frac{d}{HolePitch}\right)^{0.1} = 0.98\left(\frac{6.4}{35.355}\right)^{0.1} = 0.826 \qquad \textbf{(3-22)}$$

Finally, combining Equations 3-13, 3-19 and 3-22, C_2 was calculated:

$$C_2 = \frac{1}{(0.826)^2 \times 0.02}\left[\left(\frac{100}{5}\right)^2 - 1\right] \approx 29240\ m^{-1} \qquad \textbf{(3-23)}$$

By adjusting C_2 in the x-direction, the resistance of the baffles to flow in x-direction was taken into account. As recommended in the Fluent 6.3 user manual, for convergence purposes, a factor of 1000 is used for the constant C_2 in the y and z directions if the resistance to flow in y and z directions through the baffles is much higher than that in the x-direction. Therefore, C_2 was set to $2.924 \times 10^7\ m^{-1}$ for flow in y and z directions through baffles.

3.3.3.2 Modeling the Wire Mesh Demister

As noted in section 3.1.3.3, constants D and C_2 in the x-direction should be calculated for simulation of the flow through the wire mesh demister using the porous media model. However, the specifications of the demister were not provided in the original paper of Hansen et al. (1993). Therefore, the most commonly used wire mesh properties were assumed for calculation of these constants. A wire mesh pad with a thickness of 0.15 m is commonly used in separators (Walas, 1990; Lyons and Plisga, 2005; Coker, 2007), hence a type E as specified by Helsør and Svendsen (2007) was selected for simulation purposes. Specifications of the selected mesh pad as well as correlating parameters were taken from Helsør and Svendsen (2007) and are presented in Table 3-10.

Table 3-10. The physical properties and correlation parameters given for wire mesh pad of type E (Helsør and Svendsen, 2007).

Thickness (m)	Wire Diameter (mm)	Mesh Density (kg/m^3)	Mesh Specific Surface Area (m^2/m^3)	Porosity (%)	α (m^2)	C (m^{-1})
0.15	0.27	186.9	345	97.7	2.6×10^{-7}	63

Based on the parameters reported in Table 3-10 and using Equation 3-15, the constants D and C_2 in the x-direction were calculated as follows: $D = 3.85\times10^6\ m^{-2}$; $C_2 = 126\ m^{-1}$. Again, as was the case when adjusting parameters in y and z directions for the baffles, a factor of 1000 was used for constants D and C_2 in the y and z directions: $D = 3.85\times10^9\ m^{-2}$; $C_2 = 1.26\times10^5\ m^{-1}$.

3.3.3.3 Developed Model for Physical Verification

Although the best parameter estimates, based on experience, were made while modeling the baffles and demister via the porous media model in Fluent, the upper section of separator, in which the internals were installed, was simulated to make certain that the pressure and velocity profiles were of an expected shape[9]. The steps for mesh generation are the same as those previously described for the whole separator. Inlet gas velocity and outlet pressure of 6870 kPa were input as the boundary conditions, and the SIMPLE method with the following settings was used as the solver:

Discretization Method for Pressure: Standard
Discretization Method for Momentum: Second Order Upwind
Under-Relaxation Factor for Pressure = 0.7
Under-Relaxation Factor for Density = 0.7
Under-Relaxation Factor for Body Force = 0.7
Under-Relaxation Factor for Momentum = 0.0001
Under-Relaxation Factor for Turbulent Groups (Kinetic Energy, Dissipation Rate, and Viscosity) = 0.7

[9] The results will be presented in Chapter Four.

3.3.4 *Definition of Droplet Size Distribution*

In order to model the dispersion of oil and water droplets in the fluid flow domain, the specification of particle size distribution is a key step. Any relevant experimental data would have been very useful. However, this crucial information was not given in the paper of Hansen et al. (1993). Thus, a reliable method was required for prediction of particle size distribution for oil and water droplets entering the separator. There are numerous research studies that predict the size distribution of fluid dispersions. However, most of them have focused on prediction of maximum stable droplet size. Because, the other necessary size distribution parameters such as spread parameter, minimum and mean droplet size can be estimated based on the predicted (or measured) maximum stable droplet size and the nature of the fluid phases. In the present study, the Rosin-Rammler (1933) particle size distribution function (Equation 3-16) has been used. In Equation 3-16, the volume mean diameter, $\bar{d}$, can be estimated from maximum droplet diameter, d_{max}, via Equation 3-24 (Green and Perry, 2008):

$$\bar{d} = 0.4\ d_{\max} \qquad \textbf{(3-24)}$$

Furthermore, to specify the spread parameter, n, for the Gullfaks-A dispersions, two experimental studies, performed by Karabelas (1978) and Angeli and Hewitt (2000), were used. The experiments of Karabelas were carried out with kerosene (ρ = 798 kg/m^3, μ = 0.00182 $Pa.s$) and a more viscous transformer oil (ρ = 892 kg/m^3, $\mu \cong$ 0.0156 $Pa.s$) as continuous phases and water as dispersed phase. The experiments of Angeli and Hewitt were performed with both water and the oil (ρ = 801 kg/m^3, μ = 0.0016 $Pa.s$) as dispersed and/or continuous phases. The experimental distributions of Angeli and Hewitt produced a value between 2.1 and 2.8 for the Rosin-Rammler spread parameter. This result agrees with the values of 2.13 to 3.30 reported by Karabelas (1978) for water dispersed in two different oils. As one of the most interesting experimental results, Karabelas (1978) emphasized that the spread parameter can be assumed to be constant and close to their measured average value while dealing with oil in water or water in oil dispersions. Therefore, the arithmetic average value of 2.6, as reported by Karabelas (1978), was used while setting the particle size distribution in this study.

The next step involved finding a reliable method for prediction of maximum stable oil and water droplet sizes. As noted, many methods have been presented. The fundamental theoretical work in the field of droplet dispersions in the turbulent flow was conducted independently by Kolmogorov (1949) and Hinze (1955). They assumed that the maximum stable droplet or bubble size, d_{max}, could be determined by the balance between the turbulent pressure fluctuations, tending to deform or break the droplet or bubble, and the surface tension force resisting any deformation. In their developed theory, the ratio of these forces was included through the critical Weber number, Equation 3-25:

$$We_{crit} = \frac{\tau}{\sigma / d_{\max}} \qquad \textbf{(3-25)}$$

where τ is dynamic pressure fluctuations in *Pa*, and σ is surface tension in *N/m*. Equation 3-25 can be combined with the equations for turbulent fluid flow in a pipeline to calculate the maximum stable droplet or bubble size. In this regard, Equation 3-26 calculates the dynamic pressure fluctuations:

$$\tau = 2\rho_c \left(\psi d_{\max}\right)^{2/3} \qquad \textbf{(3-26)}$$

where ψ is energy dissipation rate per unit mass in *W/kg* which can be calculated by Equation 3-27:

$$\psi = \frac{2 f V_c^{\,3}}{D} \qquad \textbf{(3-27)}$$

where f is friction factor, V_c is superficial velocity of continuous phase in *m/s*, and D is inside diameter of pipe in *m*. The friction factor can be calculated using the Blasius equation, Equation 3-28:

$$f = 0.079\,\mathrm{Re}^{-0.25} \qquad \textbf{(3-28)}$$

By substitution of Equations 3-26 to 3-28 into Equation 3-25, $d_{\max}$ can then be estimated using Equation 3-29:

$$d_{\max} = 1.38\left(We_{crit}\right)^{0.6}\left(\frac{\sigma^{0.6}}{\rho_c^{\ 0.5}\mu_c^{\ 0.1}}\right)\left(\frac{D^{0.5}}{V_c^{\ 1.1}}\right) \qquad \textbf{(3-29)}$$

where ρ_c and μ_c are the continuous phase density in *kg/m*^{3} and viscosity in *Pa.s*, respectively.

The other important theory for maximum stable bubble size was developed by Levich (1962). He assumed that the maximum stable droplet or bubble size $d_{\max}$, could be determined by the balance between the internal pressure of the droplet or bubble and the capillary pressure of the deformed droplet or bubble. In his developed theory, the ratio of these forces was taken into account by the critical Weber number in a new form, Equation 3-30:

$$We'_{crit} = \frac{\tau}{\sigma/d_{\max}}\left(\frac{\rho_d}{\rho_c}\right)^{1/3} \qquad \textbf{(3-30)}$$

Similar to the Kolmogorov-Hinze theory (Kolmogorov, 1949; Hinze, 1955), Equation 3-30 can be combined with Equations 3-26 through 3-28 to calculate the maximum stable droplet or bubble size $d_{\max}$, Equation 3-31:

$$d_{\max} = 1.38\left(We'_{crit}\right)^{0.6}\left(\frac{\sigma^{0.6}}{\rho_c^{\ 0.3}\rho_d^{\ 0.2}\mu_c^{\ 0.1}}\right)\left(\frac{D^{0.5}}{V_c^{\ 1.1}}\right) \qquad \textbf{(3-31)}$$

where ρ_d is dispersed phase density in *kg/m*^{3}, and the other parameters were defined while deriving Equation 3-29.

In an excellent study, Hesketh et al. (1987) modified the Levich theory to develop an equation that includes all the salient physical fluid properties required to describe droplet or bubble size in turbulent flow. Hesketh et al. (1987) considered both the Kolmogorov-Hinze and Levich theories and recognized that in predicting maximum particle size for liquid-liquid and gas-liquid dispersions, only the latter gives consistent results. The critical Weber number should depend only on the mechanism of breakup, not on the fluid physical properties. For the liquid-liquid dispersions, the values of We_{crit} calculated from the Kolmogorov-Hinze theory are some 10% of those calculated for the gas-liquid dispersions. This discrepancy implies that the breakup mechanism for liquid droplets differs from that for gas bubbles. However, the empirical findings contradict this implication, and values of We_{crit} for liquid-liquid and gas-liquid dispersions should be equal from an empirical perspective. This is what is proposed by Levich theory in that We'_{crit} for both liquid-liquid and gas-liquid dispersions are close to 1.0. Based on the Levich theory, Equation 3-31 includes all the variables necessary to describe turbulent dispersion except for the dispersed phase viscosity, μ_d. The viscous forces within a droplet or bubble increase its stability, and become important when they are of the same order of magnitude as the surface tension forces (Hesketh et al., 1987). By combining a viscosity grouping term originally proposed by Hinze (1955) with Equation 3-31, and evaluating the coefficients using experimental data, Hesketh et al. (1987) have developed the following generalized equation, Equation 3-32:

$$d_{max} = 1.38\left(\frac{\sigma^{0.6}}{\rho_c^{0.3}\rho_d^{0.2}\mu_c^{0.1}}\right)\left(\frac{D^{0.5}}{V_c^{1.1}}\right)\times$$

$$\times\left(1+0.5975\left[\frac{\mu_d\left(\mu_c^{0.25}V_c^{2.75}\rho_c^{-0.25}D^{-1.25}d_{max}\right)^{1/3}}{\sigma}\right]\sqrt{\frac{\rho_c}{\rho_d}}\right)^{0.6} \qquad \textbf{(3-32)}$$

Note that d_{max} should be calculated from Equation 3-32 in an iterative manner, since d_{max} is also present in the right hand side of Equation 3-32.

Although estimation of $d_{\max}$ using Equation 3-32 seems a bit tedious (yet not difficult if common mathematical software is used), its strong theoretical background as well as its very satisfactory representation of empirical data provided sufficient confidence in using this method for prediction of the maximum droplet size in this study. Note, Hesketh et al. (1987) showed that this approach provided excellent results when dealing with experimental data that included a broad range of physical properties: surface tension of 0.005 to 0.072 *N/m*, the continuous phase viscosity of 0.001 to 0.016 *Pa.s*, and the dispersed phase density of 1 to 1000 kg/m^3.

As mentioned, a number of attempts has been made to correlate the maximum droplet size in fluid dispersions. In order to compare the results of the selected method with those of some other available methods, in addition to the Hinze (1955) method that has been described in detail, four other correlations are as follows:

1- Sleicher (1962) correlation:

$$d_{\max} = 38\left(\frac{\sigma^{1.5}}{\rho_c V_c^{2.5} \mu_c^{0.5}}\right)\left[1+0.7\left(\frac{\mu_d V_c}{\sigma}\right)^{0.7}\right] \quad \textbf{(3-33)}$$

2- Tatterson et al. (1977) correlation:

$$d_{\max} = 0.265\left(\frac{\sigma^{0.5} D^{0.6}}{\rho_c^{0.4} V_c^{0.9} \mu_c^{0.1}}\right) \quad \textbf{(3-34)}$$

Note, the original equation, with the coefficient of 0.106 on the right hand side, was proposed for estimation of volume median droplet size. The presented form is based on Equation 3-24.

3- Karabelas (1978) correlation:

$$d_{\max} = 4.0\left(\frac{\sigma^{0.6} D^{0.4}}{\rho_c^{0.6} V_c^{1.2}}\right) \quad \textbf{(3-35)}$$

4- Angeli and Hewitt (2000) correlation:

$$d_{max} = \frac{4.2\times10^{-8}}{V_c^{1.8} f^{3.13}} \quad \textbf{(3-36)}$$

Note, all the variables present in these correlations have been defined while developing Equation 3-32. The maximum stable droplet sizes calculated by various methods or correlations for dispersions of Gullfaks-A separator are reported in Table 3-11. Assuming that Equation 3-32 predicts d_{max} for the Gullfaks-A dispersions with the best reliability, the other approaches can be sorted (in a descending order) based on their average accuracies: 1- Tatterson et al. (1977), 2- Karabelas (1978), 3- Hinze (1955), 4- Sleicher (1962), 5- Angeli and Hewitt (2000).

Based on the discussed procedures, all the discrete phase parameters necessary for CFD simulation of Gullfaks-A separator have been calculated and presented in Table 3-12.

Table 3-11. The maximum stable droplet sizes calculated by various methods for dispersions of Gullfaks-A separator.

Method	Calculated Values for 1988 Condition (*m*)		Calculated Values for the Future Conditions (*m*)	
	Oil Drops	Water Drops	Oil Drops	Water Drops
Hinze (1955)	0.004345	0.008071	0.003744	0.006954
Sleicher (1962)	0.01799	0.04924	0.01374	0.03632
Tatterson et al. (1977)	0.003434	0.005753	0.003071	0.005145
Karabelas (1978)	0.003615	0.006715	0.003115	0.005785
Hesketh et al. (1987)	0.002267	0.004006	0.001955	0.003452
Angeli and Hewitt (2000)	0.03857	0.03857	0.03088	0.03088

Table 3-12. The discrete phase parameters used in CFD simulation of Gullfaks-A separator.

Discrete Phase Parameters	1988 Condition		Future Condition	
	Oil Drops	Water Drops	Oil Drops	Water Drops
Maximum Diameter (μm)	2267	4000	1955	3450
Mean Diameter (μm)	907	1600	780	1380
Total Mass Flow-rate (kg/s)	6.5×10^{-4}	4.4×10^{-3}	4.2×10^{-4}	2.8×10^{-3}
Number of Tracked Particles	1000			
Minimum Diameter (μm)	100			
Spread Parameter	2.6			

3.3.5 *Setting CFD Simulator Parameters*

After importing the mesh file and making the modification to it, the necessary material properties for various phases were input. Mesh modification consisted of converting the highly skewed grids to polyhedral grids which resulted in a minor reduction in the number of cells. Then, since the Reynolds number was much more than the transient value (Re = 2300) for all fluid phases, the common turbulent flow model of $k-\varepsilon$ was selected.

In order to set the boundary conditions for inlet, the velocity and volume fractions of phases were set. For the gas-outlet boundary, outlet pressure and volume fractions (as pure gas) were set while for the liquid-outlet boundaries, outlet velocities and volume fractions (as pure liquid) were set. For setting the flow regimes in inlet and outlet nozzles, the turbulence intensity and hydraulic diameter of flow through the nozzles were determined. Similar to the CFD modeling of the two-phase separators, discussed in section 3.2, the turbulence intensity in the inlet and outlet zones was estimated using Equation 3-17. Therefore, the boundary condition inputs could be calculated based on the separator dimensions and various phase properties. Table 3-13 provides the calculated input values. Note, that in setting the gas outlet boundary condition, the pressure outlet was set to a constant value of 6870 *kPa*.

For the dispersed liquid droplets the momentum reduction of droplets after hitting the inside walls of vessel, deflector baffle, splash plate, and weir was set. In the current study, it was assumed that droplets lose 90% of their momentum after coming into contact with the solid surfaces. Thus, normal and tangent reflection coefficients were set to be 0.10 for all the solid surfaces.

Table 3-13. The calculated boundary condition inputs used in CFD simulation of Gullfaks-A separator.

	1988 Velocity (*m/s*)	Future Velocity (*m/s*)	*I* (1988)	*I* (Future)
Inlet Mixture	5.8788	6.6560	2.90%	2.73%
Gas Outlet	——	——	2.28%	2.28%
Oil Outlet	4.4359	3.3293	3.35%	3.48%
Water Outlet	0.6919	2.9991	3.01%	2.51%

During normal operation, the separator was half-filled with liquid (Hansen et al., 1993). Furthermore, based on the "horizontal separator with weir" design procedure (Monnery and Svrcek, 1994), the normal liquid level for the heavier phase is typically set to half of the weir height. Therefore, the liquid levels for oil and water phases were set to be 1.664 *m* and 0.625 *m*, respectively. To set the position of interface between phases, the volume fractions of phases above and below the assumed interface planes were set to the reasonable values by the "Patching" tool of Fluent. Again, the iterations need to be stopped regularly, and the position of interfaces should be checked and corrected, if necessary, by "patching" the volume fractions of phases.

Based on the provided Fluent guidelines and the trend of solution convergence, the solver parameters were set as follows which did overcome stability and convergence problems:

Discretization Method for Pressure: Body Force Weighted
Discretization Method for Momentum and Volume Fraction: First Order Upwind
Solution Method: PISO
Under-Relaxation Factor for Pressure = 0.1
Under-Relaxation Factor for Density = 0.9
Under-Relaxation Factor for Body Force = 0.9
Under-Relaxation Factor for Momentum = 0.0005
Under-Relaxation Factor for Volume Fraction = 0.005
Under-Relaxation Factor for Turbulent Groups (Kinetic Energy, Dissipation Rate, and Viscosity) = 0.7

3.4 Summary

The steps required for CFD simulation of multiphase separators have been presented. First, the pertinent concepts of CFD modeling were introduced, and then the requirements for the use of Fluent 6.3.26 for the simulation of multiphase separators were described. Two cases, i.e. pilot-plant-scale two-phase separators and the large-scale three-phase separator, were CFD simulated. Two approaches, Discrete Phase Model (DPM) and a combination of Volume of Fluid (VOF) and DPM, each with appropriate model assumptions and settings have been used. The CFD guidelines recommend the DPM approach for flow regimes in which the discrete phase is a small volume fraction (less than 12%), while the VOF model is recommended for the simulation of immiscible multiphase flows with specific interface surfaces. In the first case study, both the DPM and the VOF-DPM approaches were tested. However, based on the results obtained from this case study, only the VOF-DPM approach was used for simulation of the complex features of three-phase separation in the large-scale separator. To implement the combined VOF-DPM approach, having developed a CFD simulation of the overall phase behavior of the fluid flows using the VOF model, droplets of oil and water were injected through the inlet nozzle to be tracked by DPM model. Consequently, the VOF model was used to provide the overall picture of fluid flow behavior in the separators, and the DPM model was used to track the liquid droplets injected, moving the simulation toward a realistic situation.

The distribution baffles and mist eliminator were modeled using the Porous Media Model of Fluent. The use of the Porous Media Model required utilization of the available specifications and design information for three-phase separators.

By exploiting the available theoretical approaches and experimental correlations, a useful methodology for estimation of particle size distribution, which is necessary for implementing DPM approach, has been developed.

Chapter Four: Results and Discussion

The CFD simulation results for the pilot-plant scale two-phase separators and the large-scale three-phase separator will be presented. For the two-phase separators, two simulation approaches, DPM and VOF-DPM, were used. Based on the results for the two-phase separators, only the VOF-DPM modeling approach was implemented for simulation of the three-phase separator.

4.1 Simulation of the Pilot-Plant-Scale Two-Phase Separators

Two approaches were implemented for the simulation of the pilot-plant-scale two-phase separators. The first simulation used only the DPM multiphase modeling of Newton et al. (2007), and only the vapor-liquid compartment of the separators was simulated. In this approach, the gas-liquid interface was assumed as a frictionless wall which trapped the droplets coming into contact with it. In the second modeling method a combination of DPM and VOF multiphase models within Fluent 6.3.26 was used. This CFD model was based on the type of involved phase separation process and the characteristics of multiphase models of Fluent.

4.1.1 *Results of DPM Only Model*

In order to obtain the incipient velocity and separation efficiency diagram for each case, the inlet velocity was gradually increased by 0.1 m/s increments, and the flow regimes in inlet and outlet nozzles were set. The minimum velocity at which a minor fraction of the droplets (about 1wt%) were carried over in the gas phase was reported as the incipient velocity. After determining the incipient velocity, the velocity was gradually increased to 5 *m/s*, generating the separation efficiency versus velocity data. This procedure was used for the twelve cases that included four separators operating at three different pressures. Three different mean droplet sizes were also tested for each case. After setting all the CFD parameters, the number of iterations required for the continuous-phase solution convergence in vessels A, B, C, and D was 600, 500, 250, and 300, respectively. The CPU time for each iteration on a Pentium D (3.20 GHz) and 2.00 GB of RAM PC was 4, 2, 1.5, and 1 *s* for vessels A, B, C, and D, respectively. Therefore, a relatively short PC run-time of from 5 to 40 *min* was found for the solution of continuous-phase flow regimes. However, a further PC run-time of 1 to 2 *h* was required for simulation of discrete phase (liquid droplets) interaction with continuous gas phase for each case study.

4.1.1.1 Fluid Flow Profiles

The pressure and velocity profiles for the continuous phase at the incipient velocity for one of the horizontal separators, vessel C, and the vertical separator are presented in Figures 4-1 to 4-6. The attached CD contains the profiles for all the case studies. The simulated pressure profiles do indicate that the separators can be considered to operate at a constant pressure with some negligible low pressure zones in the vicinity of the gas outlet.

4.1.1.2 Droplet Coalescence and Breakup

Coalescence and breakup of droplets was also modeled and required the development of software, written in C++, that took data from the screen report of Fluent and calculated the number of droplet collisions and breakups. The results of this calculation showed that droplet coalescence at a low rate of about 0.2% was taking place. The results did show that high operating pressures and low vessel aspect ratios increased droplet coalescence.

Unlike droplet coalescence, the droplet breakup was noticeable and did show a significant variation at different operating conditions. The droplet breakup variations versus continuous phase velocity for all case studies have been presented in Appendix B. These CFD data show, as expected, that higher velocities and pressures usually increase the number of droplet breakups.

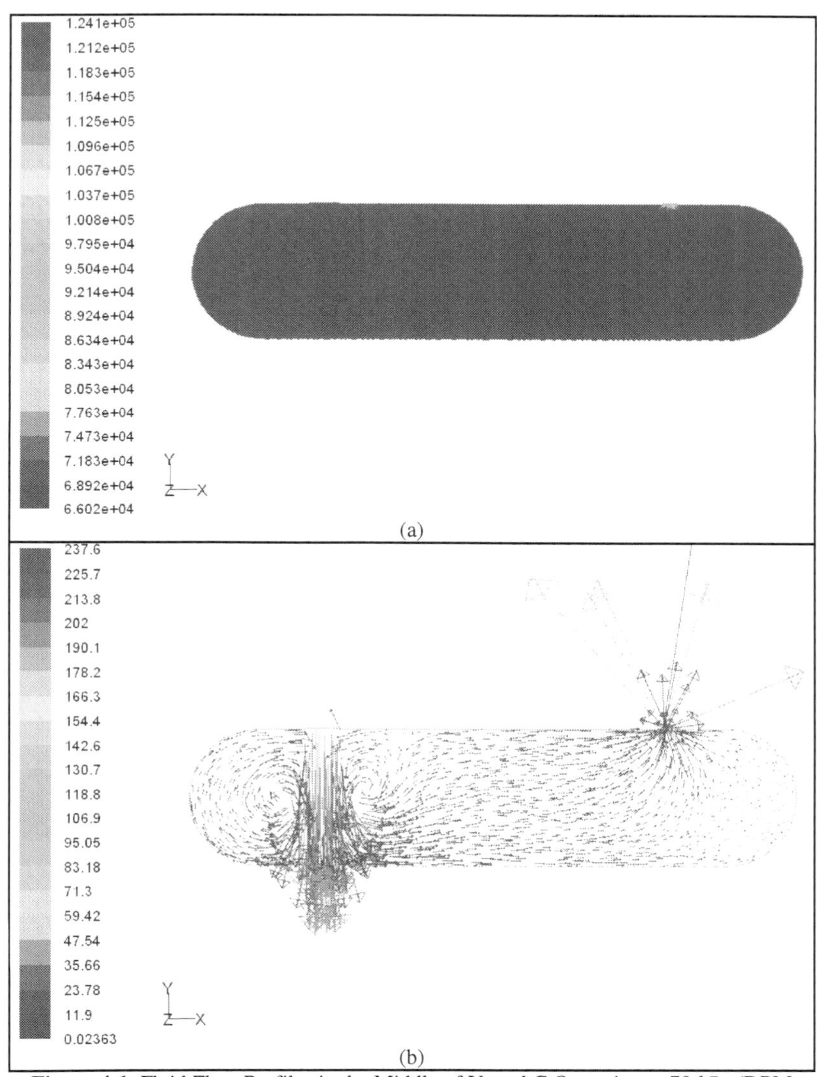

Figure 4-1. Fluid Flow Profiles in the Middle of Vessel C Operating at 70 *kPa* (DPM Approach); (a) Contours of Pressure (*Pa*), and (b) Vectors of Velocity (*m/s*).

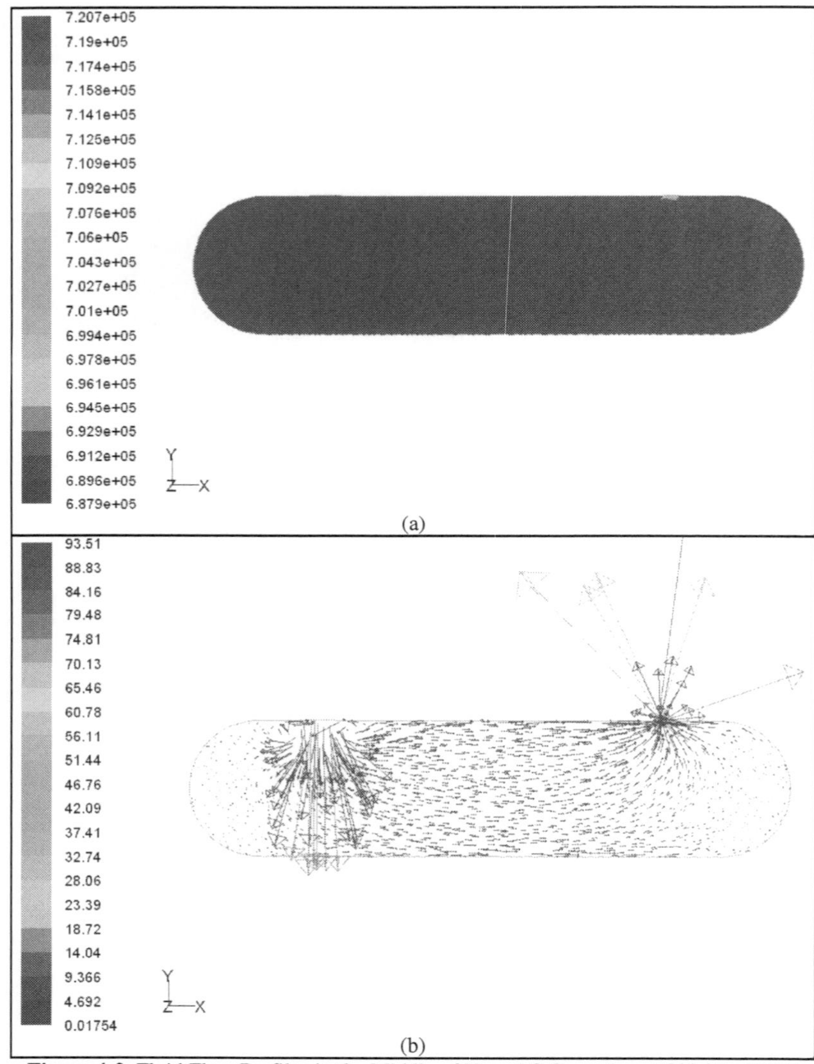

Figure 4-2. Fluid Flow Profiles in the Middle of Vessel C Operating at 700 *kPa* (DPM Approach); (a) Contours of Pressure (*Pa*), and (b) Vectors of Velocity (*m/s*).

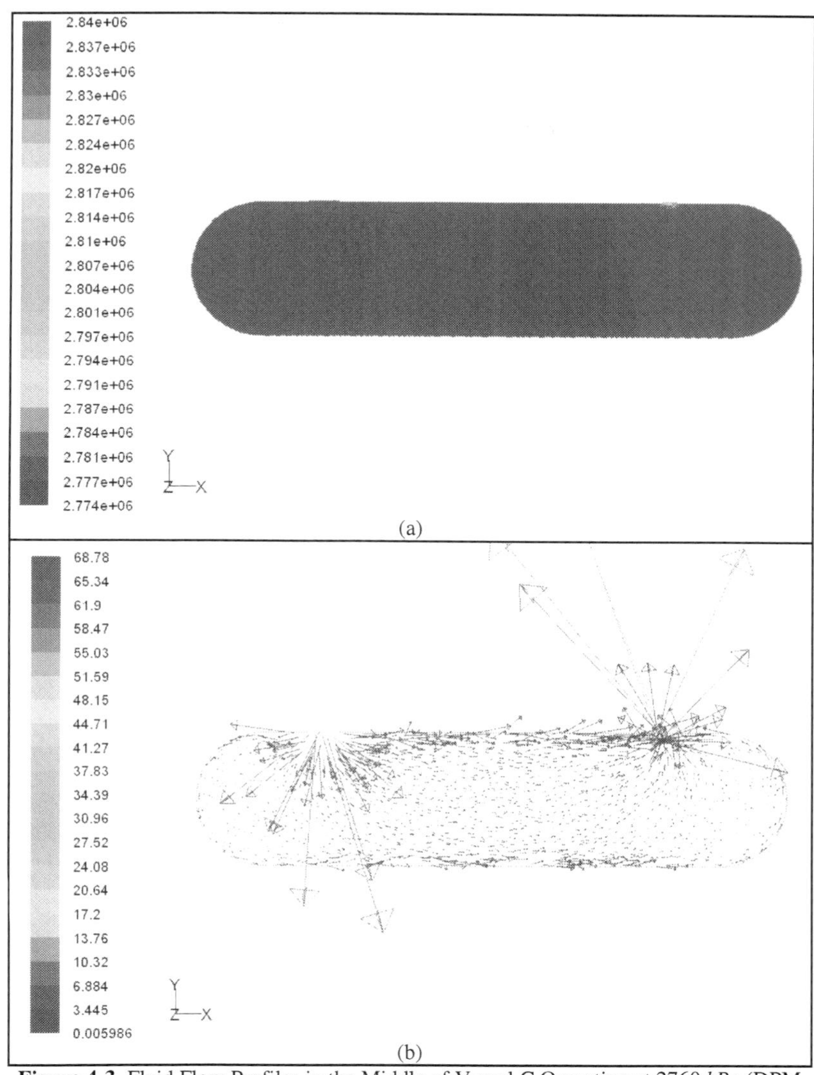

Figure 4-3. Fluid Flow Profiles in the Middle of Vessel C Operating at 2760 *kPa* (DPM Approach); (a) Contours of Pressure (*Pa*), and (b) Vectors of Velocity (*m/s*).

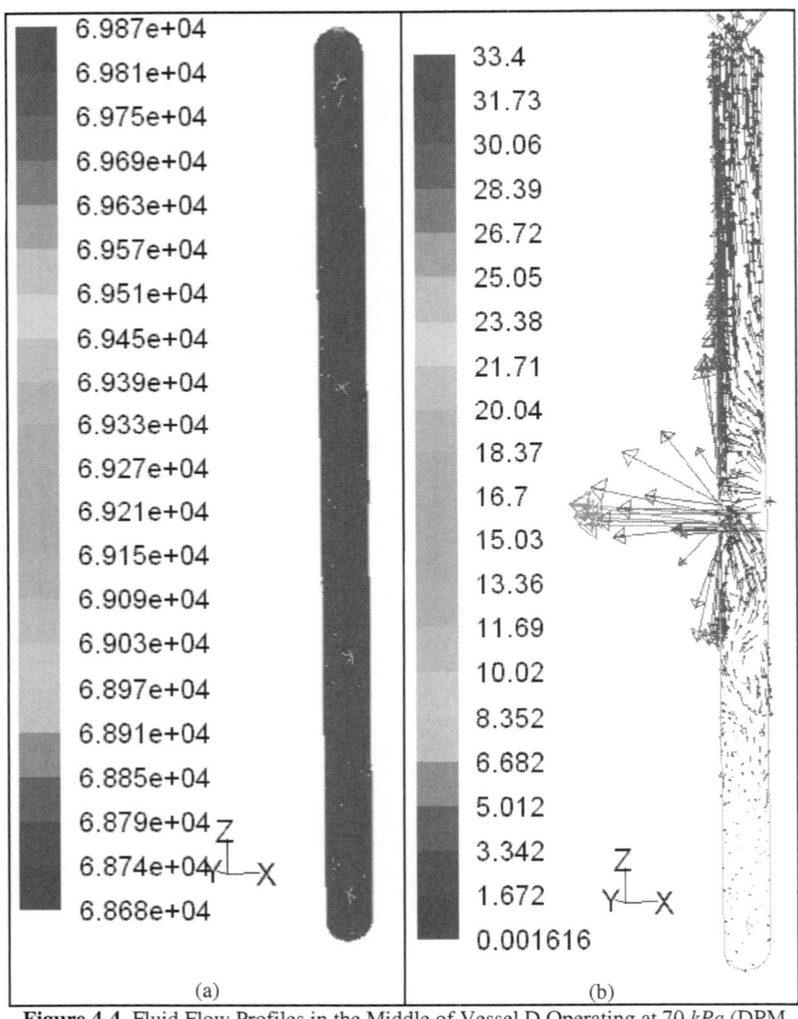

Figure 4-4. Fluid Flow Profiles in the Middle of Vessel D Operating at 70 *kPa* (DPM Approach); (a) Contours of Pressure (*Pa*), and (b) Vectors of Velocity (*m/s*).

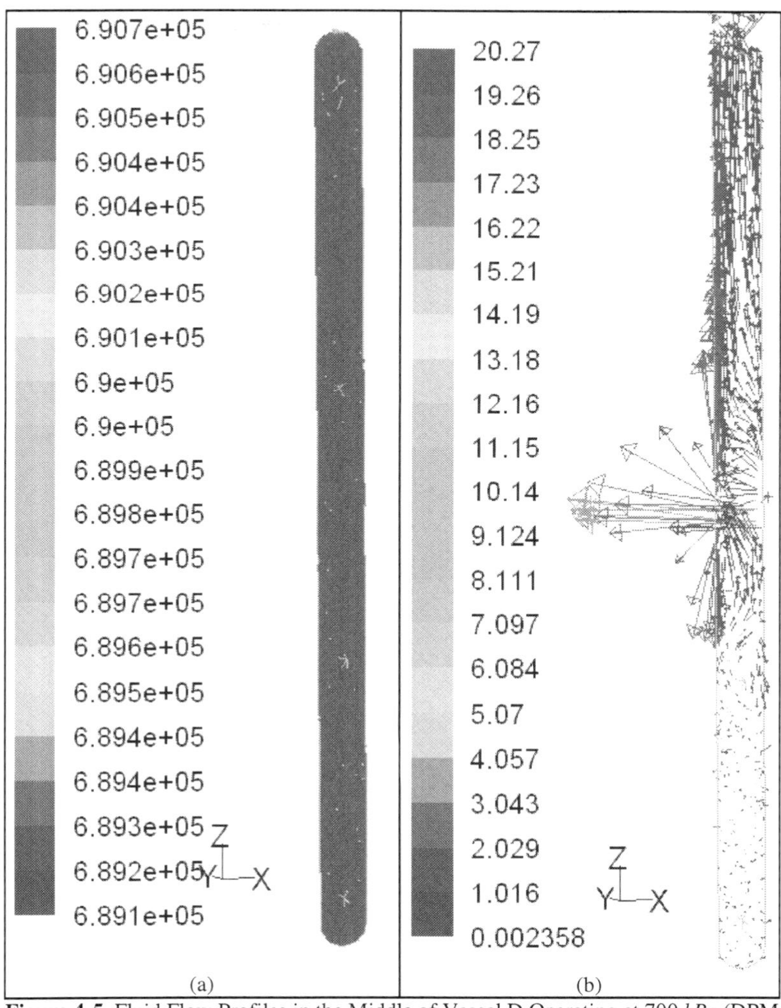

Figure 4-5. Fluid Flow Profiles in the Middle of Vessel D Operating at 700 *kPa* (DPM Approach); (a) Contours of Pressure (*Pa*), and (b) Vectors of Velocity (*m/s*).

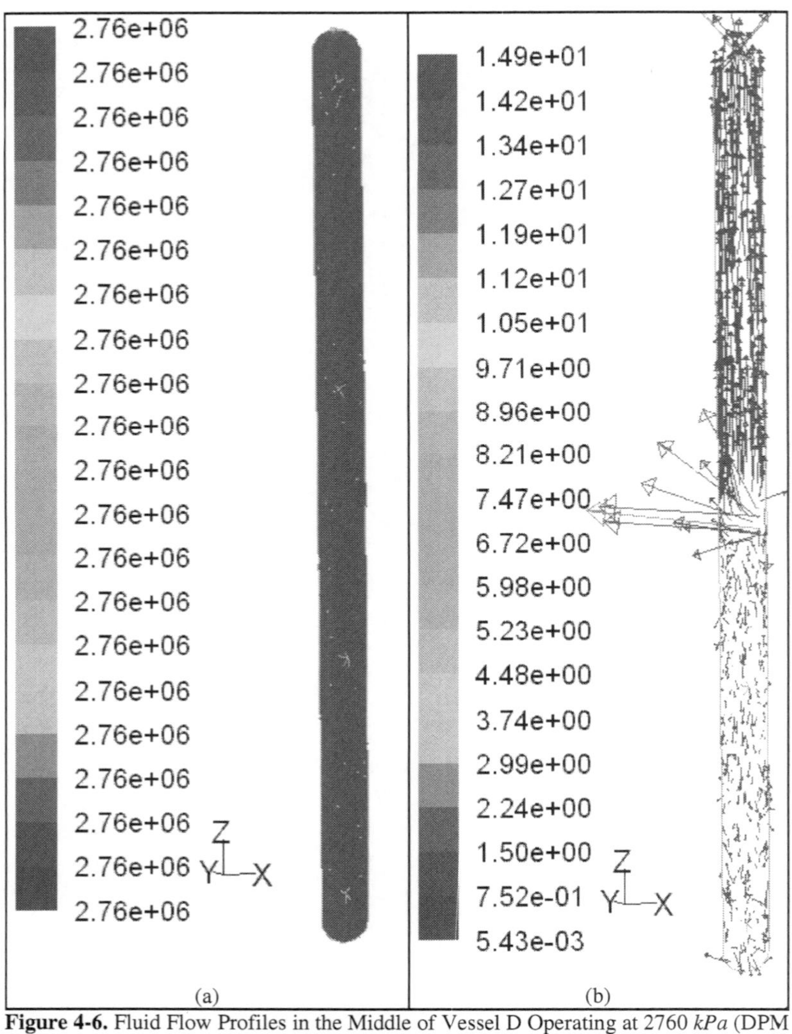

Figure 4-6. Fluid Flow Profiles in the Middle of Vessel D Operating at 2760 *kPa* (DPM Approach); (a) Contours of Pressure (*Pa*), and (b) Vectors of Velocity (*m/s*).

Table 4-1. The simulated and experimental values for incipient velocities (DPM approach).

Separator	Pressure (*kPa*)	Incipient Velocity (*m/s*)			
		Simulated			Experimental
		$\bar{d}$ = 150 μm	$\bar{d}$ = 550 μm	$\bar{d}$ = 1000 μm	
A	70	1.7	1.7	1.7	---
	700	1.0	*0.8*	*0.8*	1.0 -1.7
	2760	0.8	0.6	0.4	0.4-0.8
B	70	*1.5*	*1.4*	*1.4*	2.2-2.6
	700	0.8	0.6	0.6	0.4-1.4
	2760	0.4	0.4	0.4	0.1-0.5
C	70	2.2	*1.5*	*1.6*	2.2-2.6
	700	0.8	0.6	0.5	0.4-1.4
	2760	0.4	0.4	0.4	0.1-0.5
D	70	1.5	1.7	1.7	1.3-2.0
	700	0.8	0.9	0.9	0.7-1.7
	2760	0.6	0.6	0.6	0.4-0.8

4.1.1.3 Separation Efficiencies

Table 4-1 compares the predicted incipient velocities to the experimental values. In Table 4-1, only few of the simulated values shown in italic letters are not in good agreement with the measured experimental values. In other words, for most of the case studies, there is excellent agreement between simulated incipient velocities and the experimental data. For the droplet mean diameter of $\bar{d}$ = 150 μm, only one case study (separator B at 70 *kPa*) does not provide a good match. Therefore, one can conclude that the industry assumed mean diameter of $\bar{d}$ = 150 μm is an acceptable value for the estimation of separator incipient velocity.

Although this simple modeling approach was successful in predicting the separator incipient velocities, the resultant plots for separation efficiency versus gas velocity were not correct. A complete set of separation efficiency plots for all case studies at three different mean droplet sizes are included in the supplementary package of the book. As an example, Figure 4-7 shows such a plot for vessel C operating at 70 *kPa* and indicates that by gradually increasing the gas velocity, two zones with separation efficiencies less than 100% appear: one at about 2.2 *m/s* and the other at 2.6 *m/s*. In fact, based on the simulation results for some cases, there could be more than one of these zones, which may be referred as “carryover rush” zones. This behavior, however, is opposite to practical experience which would suggest that after reaching the incipient velocity, by increasing the velocity of continuous phase, the separation efficiency drops sharply to lower values. The other anomaly shown in this case study was a perfect separation efficiency at very high velocities (> 3.0 m/s).

These anomalies occurred in several case studies in which the DPM only model proposed by Newton et al. (2007) was used in the separator simulations. Thus, it can be concluded that modeling only the gas phase compartment of a separator does not capture the phase separation phenomenon even though the model did predict the empirical incipient velocities. The shortcoming of this very simple modeling approach is that the interactions between the liquid and gas phases and, more significantly, the dynamic interaction between the liquid droplets and continuous liquid phase are totally neglected.

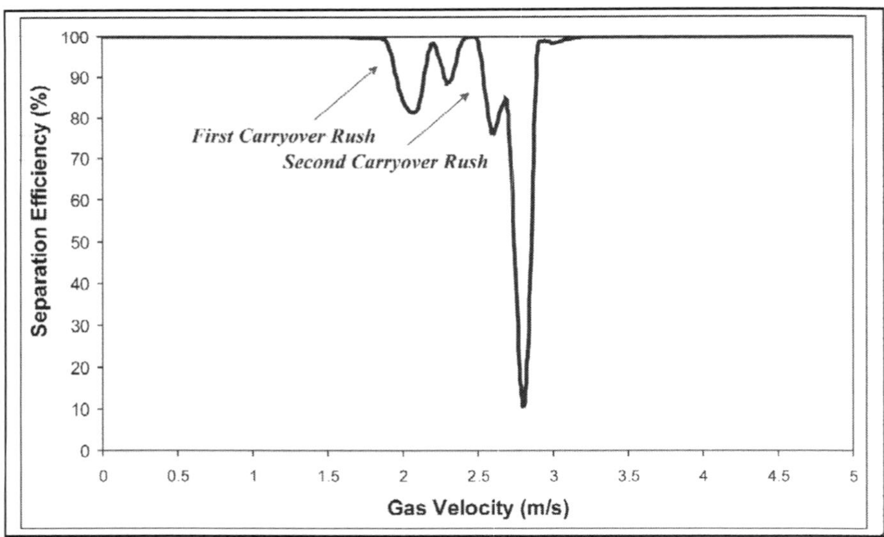

Figure 4-7. The Separation Efficiency versus Gas Velocity for Separator C at $P = 70\ kPa$ Assuming $\overline{d} = 150\ \mu m$ (DPM Approach).

4.1.2 *Results of VOF-DPM Model*

The realistic VOF-DPM model did include the continuous liquid phase in the calculations. The equations governing the injected discrete phase (liquid droplets) and coexisting continuous phases (gas and liquid phases) were solved simultaneously. This model required significantly more computational time for the calculation of the interactions among multiple discrete and continuous phases.

The number of iterations required for continuous-phase solution convergence for all the case studies was about 400. The CPU time for each iteration on a Pentium D (3.20 GHz) and 2.00 GB of RAM PC was 8, 5, 3.5, and 1 *s* for vessels A, B, C, and D, respectively. Therefore, a relatively short PC run-time of 7 to 55 *min* was sufficient for the solution of continuous-phase flow regime, but a further PC run-time of about 1 to 3 *h* was required for the simulation of interactions among multiple discrete and continuous phases.

4.1.2.1 Fluid Flow Profiles

In order to provide a macroscopic scale picture of phase separation process, the fluid flow profiles were used as the simulation output. The volume fraction contours for the case study separators operating at various pressures were essentially the same. Thus, selecting the intermediate pressure of 700 *kPa* for illustration purposes, Figure 4-8 shows volume fraction contours for all the separators. As represented, the coexisting liquid and gas phases have been separated from each other by a relatively clear interface.

The pressure and velocity profiles corresponding to the incipient velocity for one of horizontal separators, vessel B, and the vertical separator are presented in Figures 4-9 to 4-14. The supplementary package of the book includes a complete set of profile plots for all the case studies. Based on the separator pressure profiles, all the separators can be assumed to operate at a constant pressure.

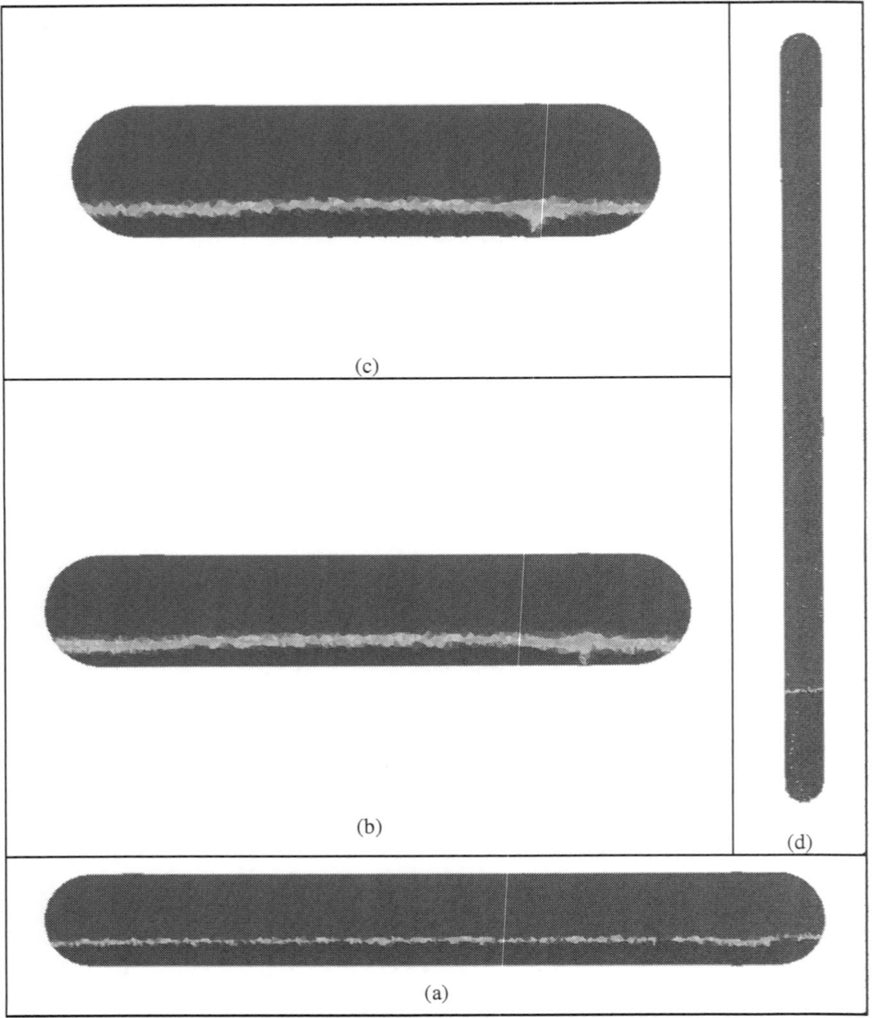

Figure 4-8. Profiles of Volume Fraction in the Middle of Two-Phase Separators; (a) Vessel A, (b) Vessel B, (c) Vessel C, and (d) Vessel D.

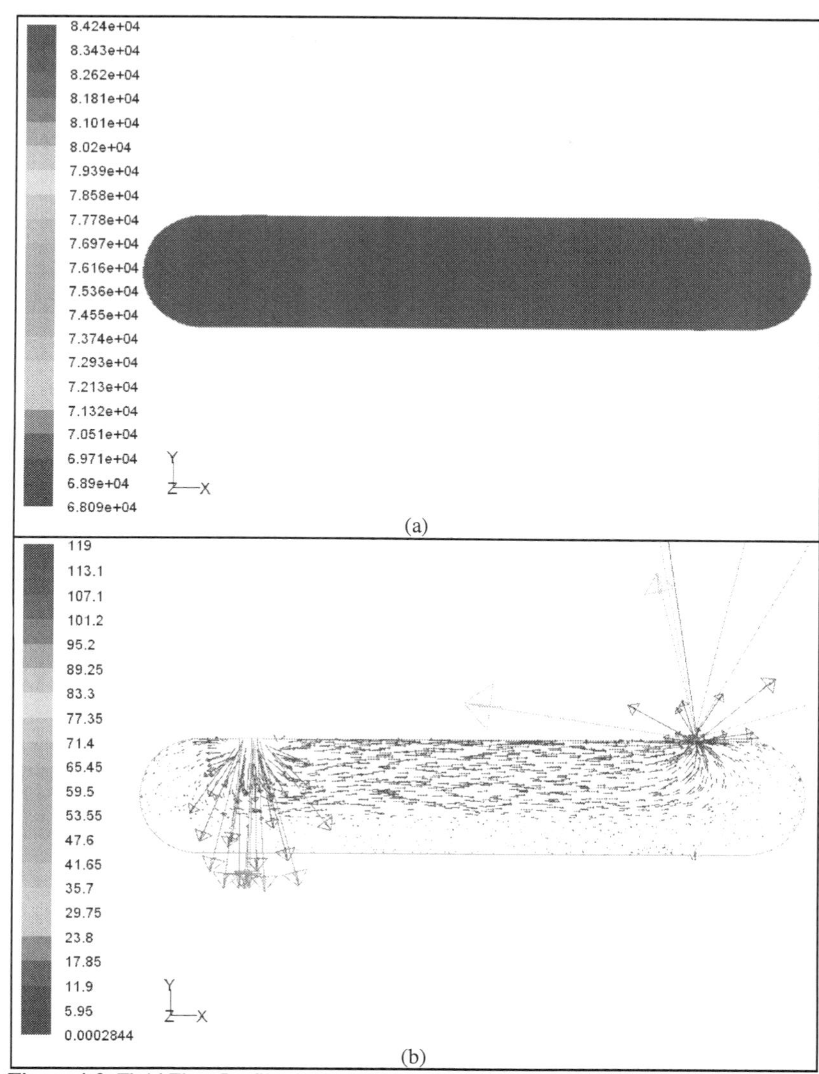

Figure 4-9. Fluid Flow Profiles in the Middle of Vessel B Operating at 70 *kPa* (VOF-DPM Approach); (a) Contours of Pressure (*Pa*), and (b) Vectors of Velocity (*m/s*).

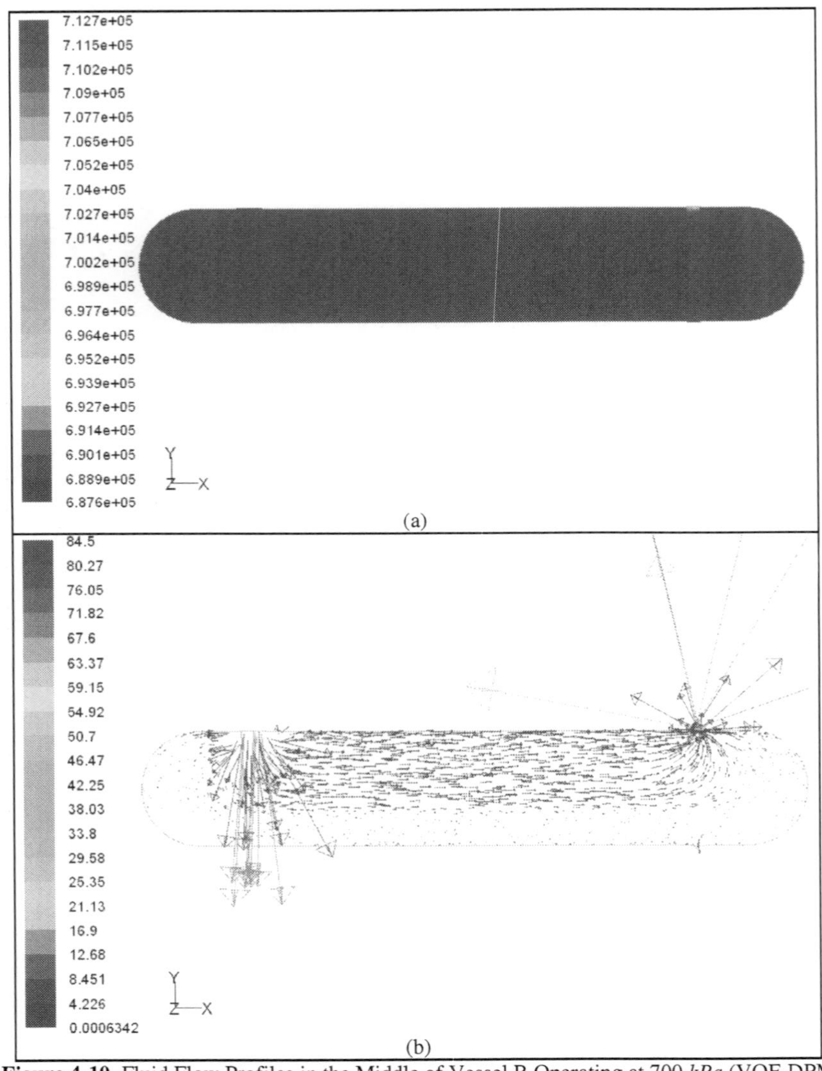

Figure 4-10. Fluid Flow Profiles in the Middle of Vessel B Operating at 700 *kPa* (VOF-DPM Approach); (a) Contours of Pressure (*Pa*), and (b) Vectors of Velocity (*m/s*).

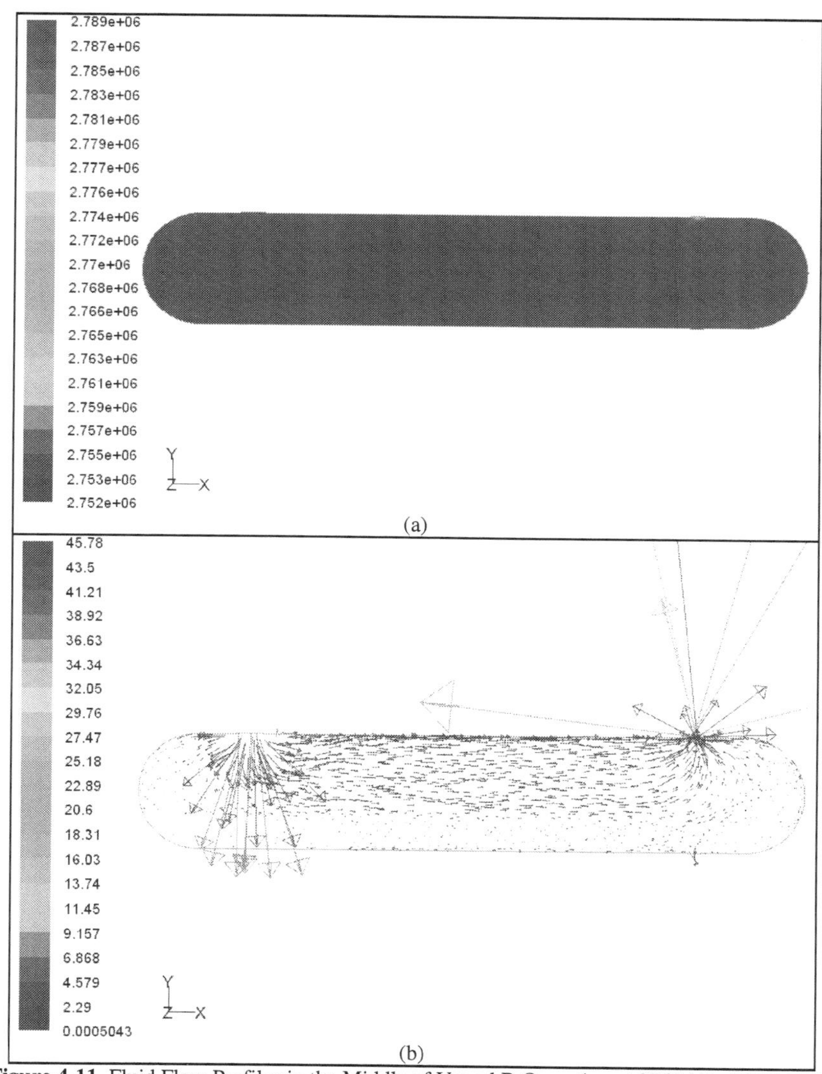

Figure 4-11. Fluid Flow Profiles in the Middle of Vessel B Operating at 2760 *kPa* (VOF-DPM Approach); (a) Contours of Pressure (*Pa*), and (b) Vectors of Velocity (*m/s*).

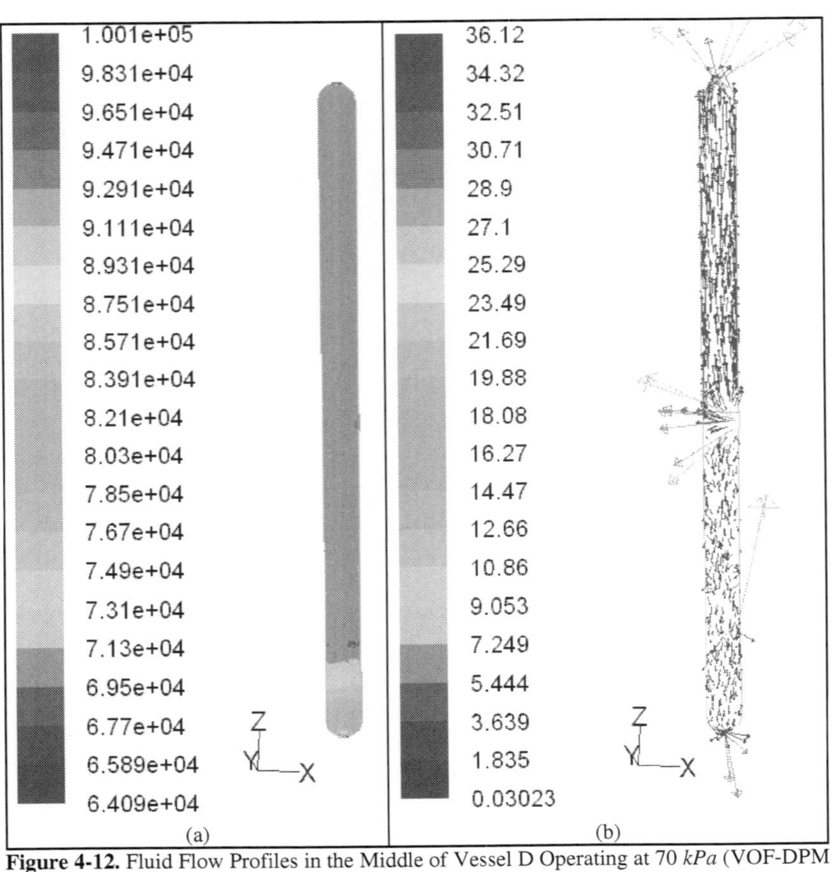

Figure 4-12. Fluid Flow Profiles in the Middle of Vessel D Operating at 70 *kPa* (VOF-DPM Approach); (a) Contours of Pressure (*Pa*), and (b) Vectors of Velocity (*m/s*).

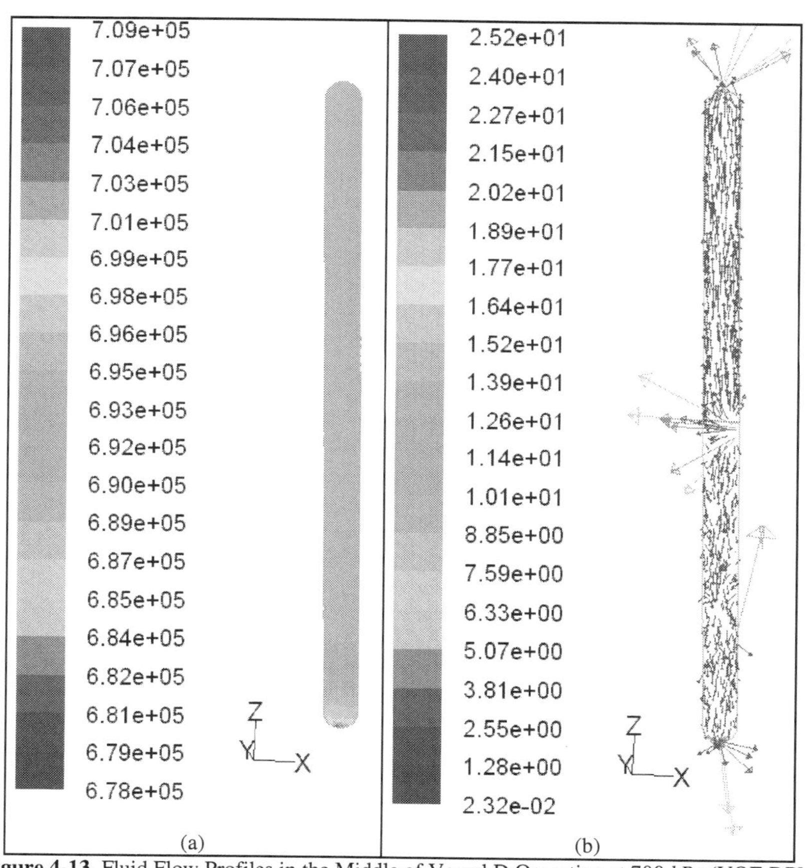

Figure 4-13. Fluid Flow Profiles in the Middle of Vessel D Operating at 700 *kPa* (VOF-DPM Approach); (a) Contours of Pressure (*Pa*), and (b) Vectors of Velocity (*m/s*).

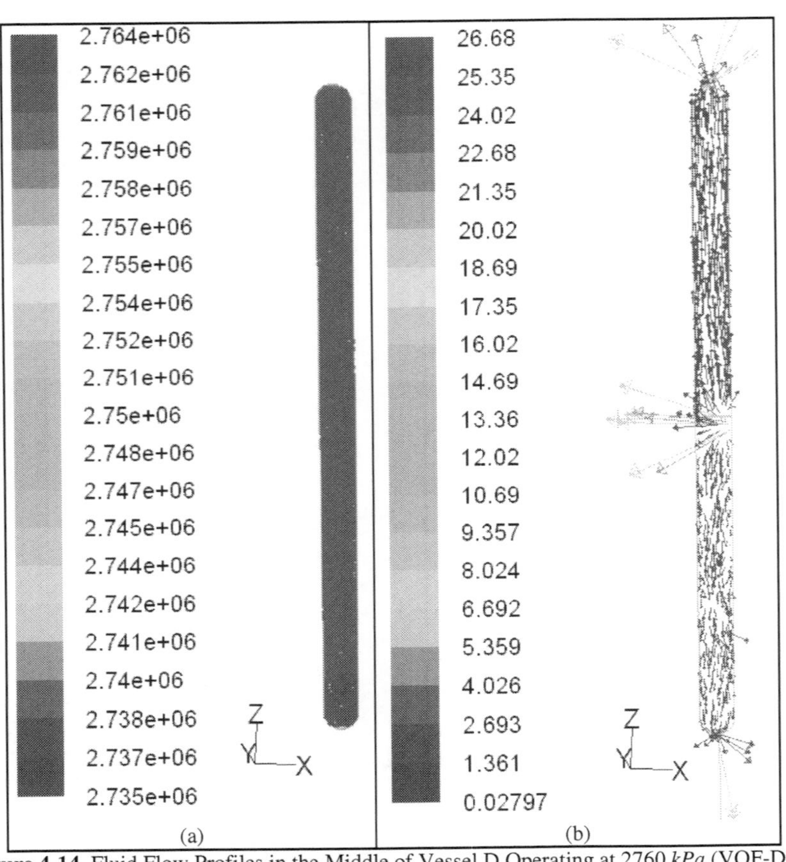

Figure 4-14. Fluid Flow Profiles in the Middle of Vessel D Operating at 2760 *kPa* (VOF-DPM Approach); (a) Contours of Pressure (*Pa*), and (b) Vectors of Velocity (*m/s*).

4.1.2.2 Droplet Coalescence and Breakup

The droplet coalescence and breakup were again modeled, and the simulation results confirmed that droplet coalescence at a rate of less than 1% was not a common phenomenon. A general conclusion for droplet coalescence could not be reached based on the simulation case study results. However, droplet breakup was a common phenomenon and did show significant variations to operating conditions. Based on the simulation results, higher velocities usually intensified the number of droplet breakups in horizontal separators, and in vertical separators, higher pressures stabilized the number of breakups to a constant rate of about 20%. Additional details on the droplet breakup variations versus operating conditions have been presented in Appendix B.

4.1.2.3 Separation Efficiencies

In order to obtain the incipient velocities and separation efficiency plots, the inlet velocity was gradually increased by 0.2 m/s increments to a maximum of 4 *m/s*. Data output surfaces were defined in the Fluent environment to enable collection of particle characteristic data as the injected droplets moved through the separator. A software program in C++ was developed to calculate the separation efficiencies based on the mass distribution of liquid droplets between the gas and liquid outlets.

The minimum velocity, at which a minor mass fraction of droplets (1wt%) were carried over in the gas phase, was assumed as the incipient velocity. The other characteristic of the incipient velocity, as also observed by Monnery and Svrcek (2000) in the pilot plant operation, was that after the incipient velocity, the carryover rate increased sharply and the separation efficiency dropped rapidly. After determining the incipient velocity, the velocity was gradually increased up to 4 *m/s* to produce the data required for separation efficiency versus velocity plots. The predicted incipient velocities and corresponding experimental values have been presented in Table 4-2. The data of Table 4-2 does show that only two cases, the values shown with italic letters, are not a good match. It is interesting to note that the predicted incipient velocity for one of these cases, separator C at 70 *kPa*, using the simple DPM model does match the experimental data quite well. Therefore, by adjusting the current model assumptions toward the DPM model assumptions, i.e. reducing the liquid level, the simulation results can be made to match the experimental data for this case.

The complete set of separation efficiency plots for all the case studies (twelve case studies) have been presented in the supplementary package of the book. However, for illustration purposes, the separation efficiency diagrams for separator B at 700 *kPa* and separator D at 70 *kPa* are shown in Figure 4-15. From a practical point of view, the obtained data curves are reasonable in that the separation efficiencies have dropped sharply after the predicted incipient velocities are exceeded. Moreover, the unacceptable results produced by the DPM simple model, i.e. perfect separation at velocities higher than the incipient velocity, have been completely eliminated. In summary, there is excellent agreement between CFD simulated phase separation behavior and the experimental data/observations for almost all of the case studies.

Table 4-2. The simulated and experimental values for incipient velocities (VOF-DPM approach).

Separator	Pressure (*kPa*)	Incipient Velocity (*m/s*)	
		Simulated	Experimental
A	70	1.6	---
	700	1.0	1.0 -1.7
	2760	0.6	0.4-0.8
B	70	*1.2*	2.2-2.6
	700	0.8	0.4-1.4
	2760	0.5	0.1-0.5
C	70	*1.2*	2.2-2.6
	700	0.6	0.4-1.4
	2760	0.5	0.1-0.5
D	70	1.5	1.3-2.0
	700	1.0	0.7-1.7
	2760	1.0	0.4-0.8

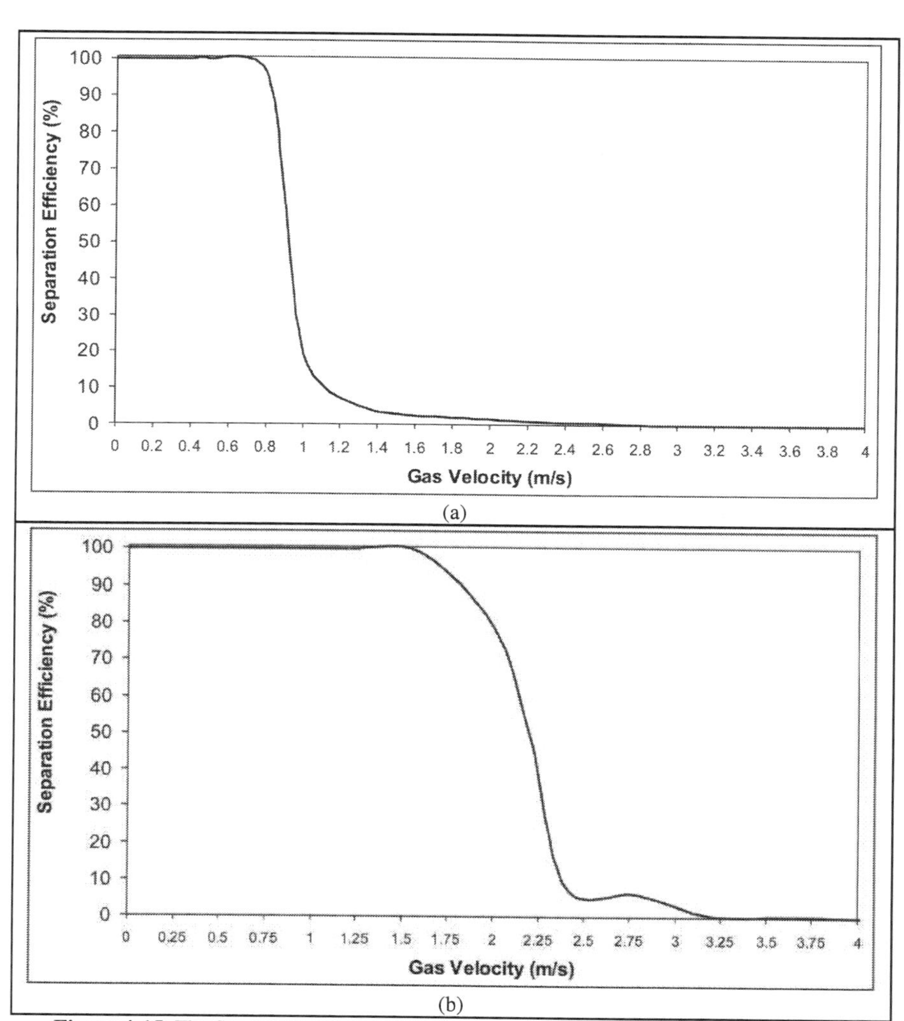

Figure 4-15. The Separation Efficiency versus Gas Velocity (VOF-DPM Approach); (a) Separator B at $P = 700\ kPa$, and (b) Separator D at $P = 70\ kPa$.

4.1.2.4 Droplet Size Distribution in Gas Outlet

Obtaining a measure of particle size distribution for the droplets in the gas outlet can also be used as a key factor in the separator performance analysis. In an efficiently designed separator, the particle size distribution in the gas-outlet zone should be very narrow and consist only of very small droplets which can be further separated by a demister. With a properly designed demister, all liquid droplets in the range of 10 to 100 μm can be removed (Smith, 1987).

A software program was developed to use the Rosin-Rammler size distribution model to calculate droplet size distribution in the gas-outlet. The results did show a variation with the gas velocity when the incipient velocity was gradually increased up to a velocity resulting in very low separation efficiency. However, only the average values of the Rosin-Rammler analysis, reported in Table 4-3, were used in assessing the separator performance. The data of Table 4-3 for horizontal separators do show that with spread parameter of about 7.1, the particle size distribution was always very narrow. Therefore, the emerging droplets were totally dominated by droplets of the mean diameter size which is always greater than 10 μm (Table 4-3). These fine droplets would hence be separated by a properly designed demister at reasonable operating condition. Provided that the load of carried over fine droplets is within the capacity of demister, high separation efficiencies may be established at velocities higher than incipient velocity.

Table 4-3. The average values of Rosin-Rammler analysis on the droplets emerging from the gas outlet (VOF-DPM approach).

Separator	Pressure (kPa)	d_{min} (μm)	d_{max} (μm)	$\bar{d}$ (μm)	n
A	70	1.27	26.11	12.11	8.34
	700	1.13	29.12	12.00	7.84
	2760	1.79	33.03	13.84	7.13
B	70	3.21	52.44	27.84	6.28
	700	5.28	49.95	30.76	7.77
	2760	3.76	34.89	16.98	6.29
C	70	1.82	37.31	18.65	7.72
	700	1.45	35.64	15.66	7.27
	2760	1.78	26.04	11.97	5.65
D	70	7.94	349.64	82.86	4.21
	700	14.88	421.79	143.39	4.60
	2760	13.79	273.67	99.15	5.06

As reported in Table 4-3, the average values of the spread parameter are generally lower for the vertical separator than those for horizontal separators, and the particle size distribution is also wider for the vertical separator. The data does show that both very small (<15 μm) and very large droplets (>>100 μm) are carried over in the vertical separator. These droplets cannot be efficiently removed by a demister and would flood it. Therefore, from a CFD point of view, separation performance of a demister for a vertical separator would be lower than that for a horizontal separator.

4.2 Simulation of a Large-Scale Three-Phase Separator

Based on the CFD simulation results obtained for the tested two-phase models and their comparison to the pilot-plant scale separator data, the combined VOF-DPM model of Fluent was used.

4.2.1 *Physical Validation of the Developed Models for the Baffles and Demister*

To physically verify the models (the Porous Media model in Fluent) used to simulate the baffles and demister, the pressure and velocity profiles in upper section of separator, in which these internals were installed, were studied. After preparing the grid system and setting all the required CFD parameters (Chapter Three), the solution converged in some 3000 iterations with a PC run-time of 10 *s/iteration*. The resultant pressure and velocity profiles are shown in Figure 4-16. Figure 4-16a does show that a low pressure drop for the gas flow through the distribution baffles and a very low pressure drop for the gas flow through the mesh demister were predicted. Therefore, the assumed geometrical specifications for baffles were validated, and the separator was shown to be operating at a constant pressure of 6870 *kPa*. Furthermore, since knitted wire mesh demisters are often causing a very low pressure drop, in the order of 250 *Pa* (Coker, 2007), the resultant negligible pressure drop is realistic. The velocity vectors shown in Figure 4-16b are reasonable and further demonstrate that the assumptions made while adjusting the porous media parameters for the distribution baffles and the wire mesh demister are realistic. Separation performance of the CFD model developed for wire mesh demister is detailed in Appendix C.

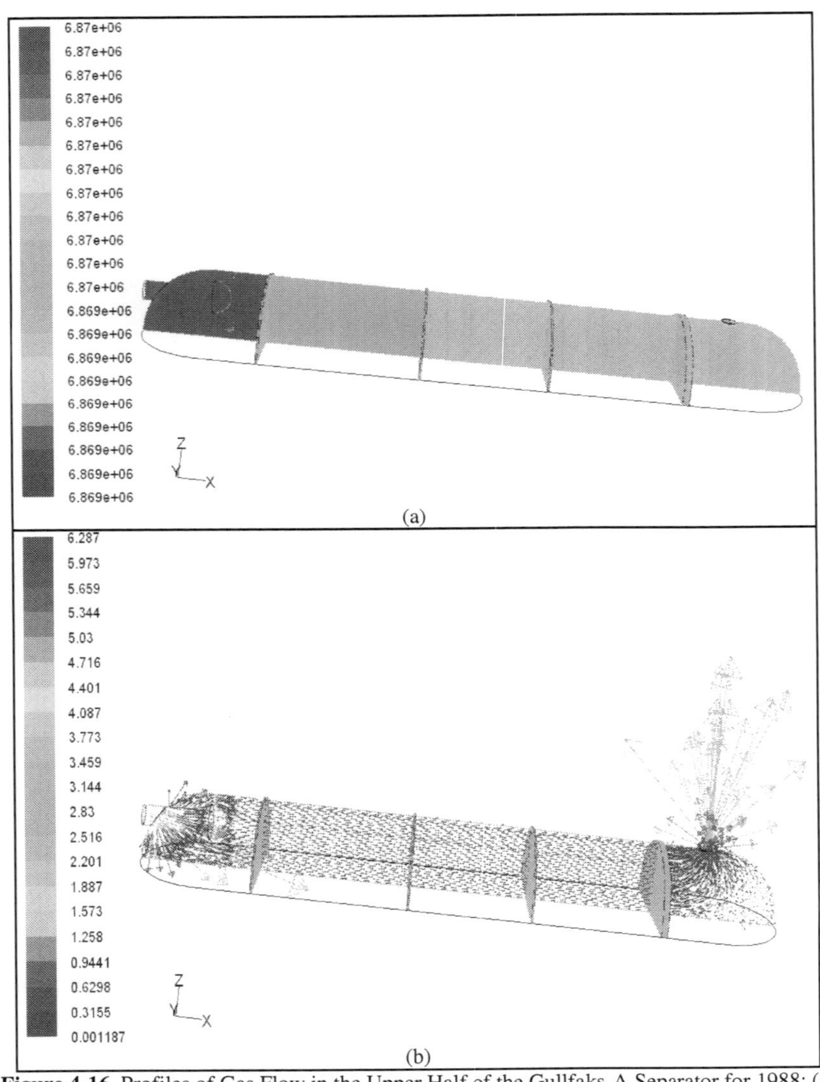

Figure 4-16. Profiles of Gas Flow in the Upper Half of the Gullfaks-A Separator for 1988; (a) Contours of Pressure (*Pa*), (b) Vectors of Velocity (*m/s*).

4.2.2 *Preliminary Considerations*

Prior to presenting the results of this case study, it is important to highlight the most important modifications of this study when compared with the original research project as presented by Hansen et al. (1993) at the 6th International Conference on Multiphase Production. In their CFD simulation of this multiphase separator, Hansen et al. (1993) made a number of simplifying assumptions as follows:

1- Fluid flow analysis was confined to the inlet zone and the bulk liquid flow zone. Therefore, the interaction between multiple zones was ignored.

2- In both zones, the flow was considered to be symmetrical around the vertical plane in the middle of the separator (xz-plane), thus, only half of each zone volume was modeled. Apparently, this is a questionable assumption, particularly, when no plug flow regime was established as shown by their results.

3- The inlet section of the separator in which all three phases are present was modeled as a two-phase gas-liquid flow, and the results did provide the boundary conditions for the distributed velocity field in the liquid pool. A two-phase simulation of a three-phase zone will reduce the accuracy of the results for the inlet zone and the incorrect boundary condition will also decrease the accuracy of downstream bulk liquid solution flow.

4- The grid systems used for numerical simulations of the inlet zone and the bulk liquid zone were $11\times8\times15$ and $23\times4\times5$, respectively. Given the vessel dimensions, the generated grid systems are rather coarse. Note, if the assumed grid system was developed such as to cover the whole vessel, the generated grid would include some 4480 cells which is 0.51% of the generated mesh cells of this study. So, the grid system of the current study is almost 200 times finer than that used in the original work of Hansen et al. (1993).

The other major improvement of this research project, when compared to not only Hansen et al. (1993) but also all the previous projects on the CFD-based study of separator performance, is the direct and quantitative evaluation of the separator efficiency. For this purpose, some data recording planes were defined to record the characteristics of the droplets passing through them. The recording surfaces of interest were two vertical yz-planes at the start and at the end of the gravity separation section, the gas outlet, the oil outlet, and the water outlet. Again, computer codes were developed to analyze the databases provided by the recording planes or the screen output of Fluent:

1- "Separation Efficiency Analyzer" dealt with databases recorded by the capturing surfaces at gas, oil and water outlets to calculate the separation efficiencies. The calculations were based on mass distribution of droplets among gas, oil and water outlets.
2- "Rosin-Rammler Analyzer" dealt with databases recorded by all the five capturing surfaces to provide the corresponding particle size distributions based on Rosin-Rammler equation.
3- "Coalescence-Breakup Analyzer" dealt with screen output of Fluent (a large transcript file) to calculate the number of droplet coalescence and breakup.

Having prepared the grid system and set all the CFD parameters, some 4000 iterations were required for the continuous-phase solution convergence. Note, the iterative process was stopped regularly to check the interface levels and correct them if necessary (Newton et al., 2007). Each iteration took about 22 *s* on a Pentium D (3.20 GHz) and 2.00 GB of RAM PC. Therefore, a PC run-time of about 24 *h* was required for solution of continuous-phase fluid flows with a further PC run-time of about 3 *h* required for the simulation of interactions among the dispersed droplets and the continuous phases.

4.2.3 *Fluid Flow Profiles*

The results of the CFD simulation of the three-phase separator in terms of fluid flow profiles are presented. The profiles of the original work of Hansen et al. (1993) have also been included (in monochrome) to make the comparison more convenient. Figure 4-17 provides the velocity vectors in the inlet zone for the 1988 production level and the predicted upcoming production conditions (referred as "future condition"). Note that the velocity vectors have not been presented for the inlet nozzle in the original work. The velocity vectors indicate the turn of inlet flow after striking the deflector baffle and the role of splash plate in stabilizing the reflected flow. Under the splash plate, an almost stagnant fluid zone can be seen. The results are comparable with those obtained in the original study of Hansen et al. (1993).

Figure 4-18 shows the velocity contours in the x-direction over the splash plate for 1988 and for the future production condition. Although the range of velocities differs from that in the original work, the overall features are similar, in that no plug flow regime is established for the inlet zone. On the splash plate, there are back flows even though the velocity distribution is more homogeneous (less back flows) for the future production condition. In the original work, back flows are predicted to occur only for the 1988 production condition.

Velocity vectors on four parallel horizontal planes (z = 0.252, 0.627, 0.999, and 1.449 *m*) and four parallel vertical planes (y = 1.4, 1.0, 0.6, and 0.2 *m*) for both 1988 and the future production condition were obtained and compared with the corresponding profiles from the original study. While Figures 4-19 and 4-20 provide some of these profiles for illustration purposes, a complete set of the profiles is in the supplementary package of the book. Using the original work profiles, Hansen et al. (1993) addressed the rotational flow regimes established between any two internals. In the present study, however, the large-scale fluid flow circulations were not present even though some minor flow circulations or back flows have been predicted. As was noted during the solution convergence trend, if the over-relaxation parameters are not adjusted correctly or the correct solver is not selected, large rotational flow patterns can be produced and the solution fluctuates without approaching a realistic converged solution.

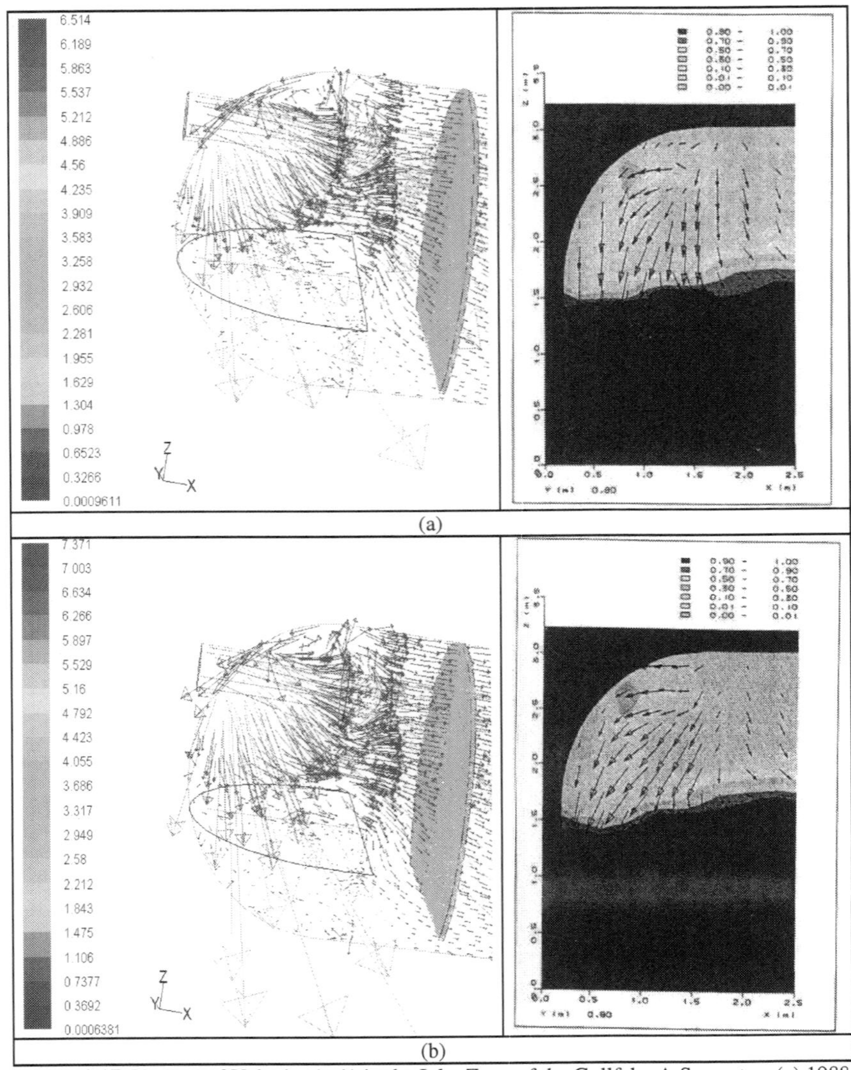

Figure 4-17. Vectors of Velocity (*m/s*) in the Inlet Zone of the Gullfaks-A Separator; (a) 1988, and (b) the Future Condition.

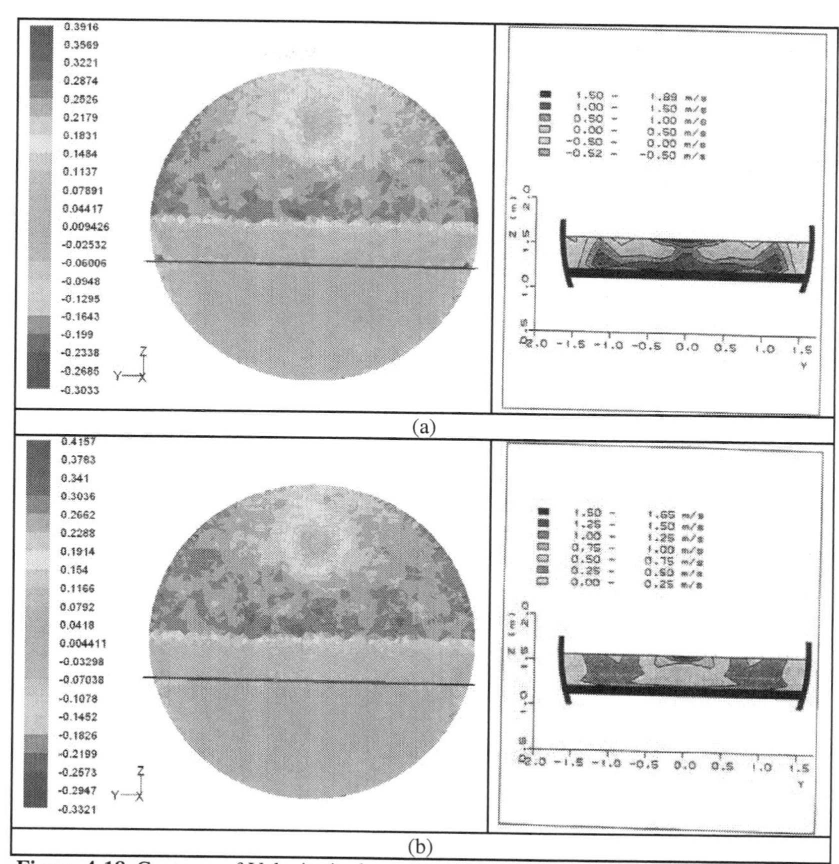

Figure 4-18. Contours of Velocity in the x-direction (*m/s*) at the End of Splash Plate of the Gullfaks-A Separator; (a) 1988, and (b) the Future Condition.

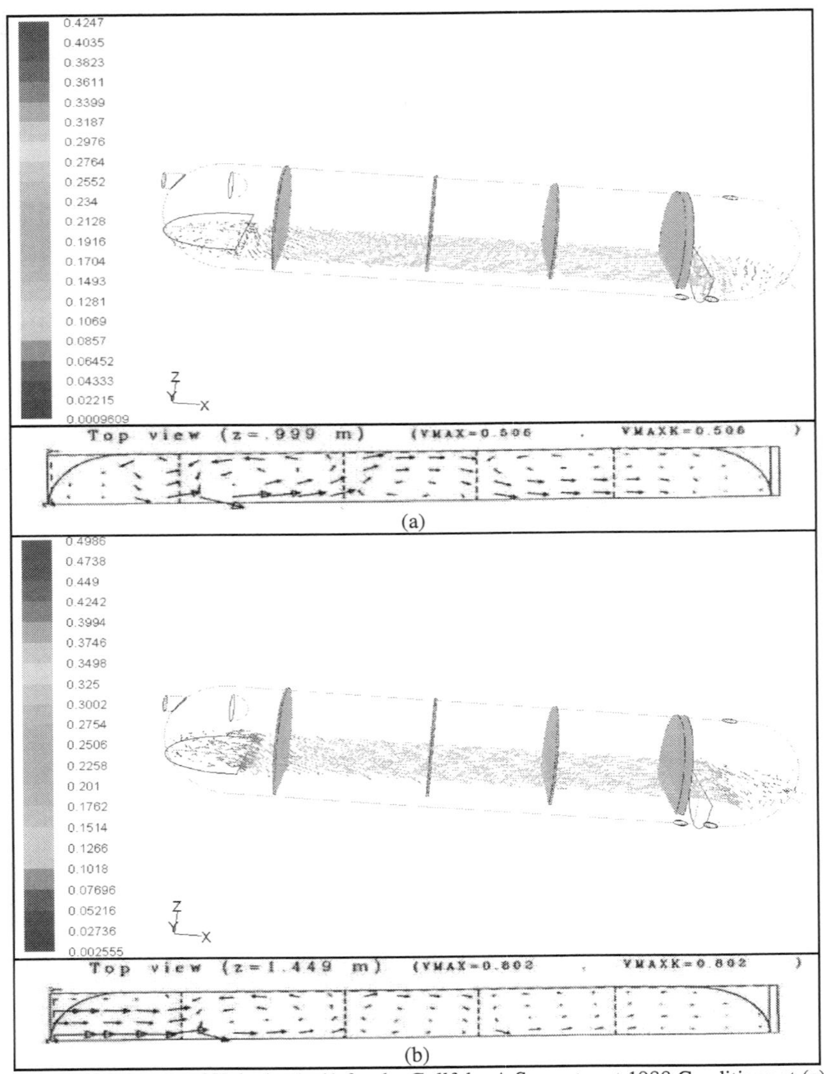

Figure 4-19. Vectors of Velocity (*m/s*) for the Gullfaks-A Separator at 1988 Conditions at (a) z = 0.999 *m*, and (b) z = 1.449 *m*.

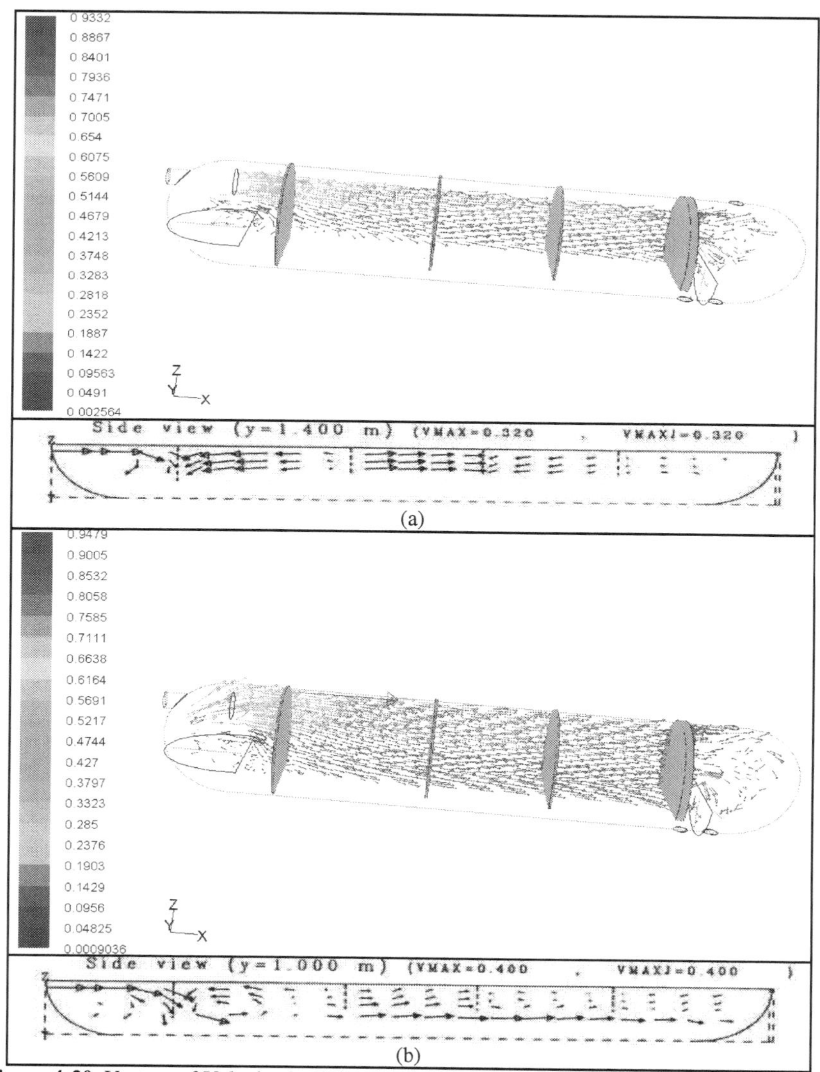

Figure 4-20. Vectors of Velocity (*m/s*) for the Gullfaks-A Separator at the Future Conditions at (a) y = 1.4 *m*, and (b) y = 1.0 *m*.

In addition to this issue, the major simplifying assumptions used in the original work are another probable source of the inaccuracy. So, it would seem that the large flow circulations predicted by Hansen et al. (1993) are a result of poor adjustments or assumptions, e.g. the poor setting of the over-relaxation parameters. Furthermore, the recent CFD-based study by Lu et al. (2007) does show that the distribution baffles generally improve the quality of liquid flow distribution in the vessel, break the large-scale circulations into smaller ones, and reduce the short-circuiting flow streams.

The pressure, velocity and density profiles for the fluid flows on the central xz-plane of the separator are shown in Figures 4-21 to 4-23. As was expected from simulation of upper section of the separator, the pressure drops assigned to the baffles and demister are small (reasonable) and the velocity vectors are also realistic. However, based on the simulated density contours, Figure 4-23, it would seem that the separator at both of operating conditions (particularly, in the future production condition) may suffer from foam and emulsion problems. The distortion of interfaces in the inlet and outlet zones does indicate a potential for foam and emulsion problems. The other detectable problem is the flow behavior near the water outlet predicted for the future production condition (Figure 4-23b). With the large increase in the produced water flow-rate, the water phase should be pumped from the vessel at much higher rates. Therefore, as indicated by the present CFD simulations, there is an increasing tendency for the oil phase to be pushed towards water outlet. This, at least, will increase the risk of turbulence in the water outlet zone and may lead to mixing of the phases. In order to overcome this problem, one should minimize the risk of mixing liquid phases by improving the vessel design. For instance, redesign of the vessel as a "bucket and weir" design may be a plausible solution.

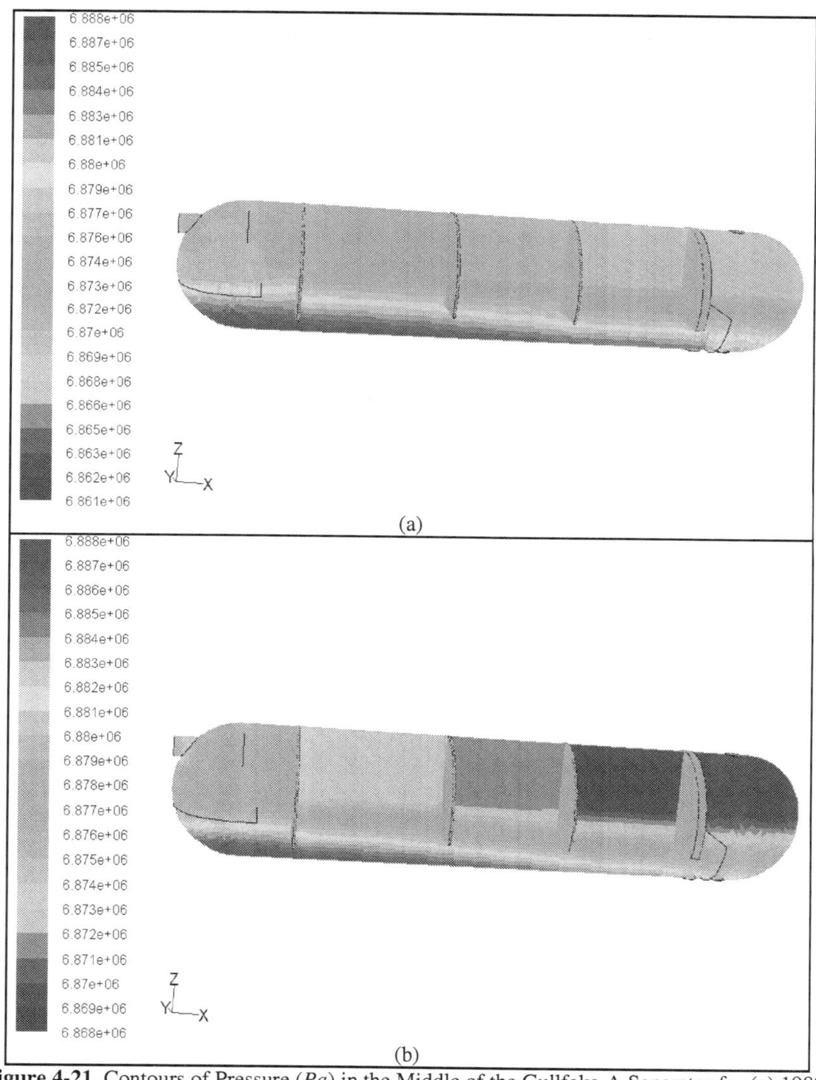

Figure 4-21. Contours of Pressure (Pa) in the Middle of the Gullfaks-A Separator for (a) 1988, and (b) the Future Condition.

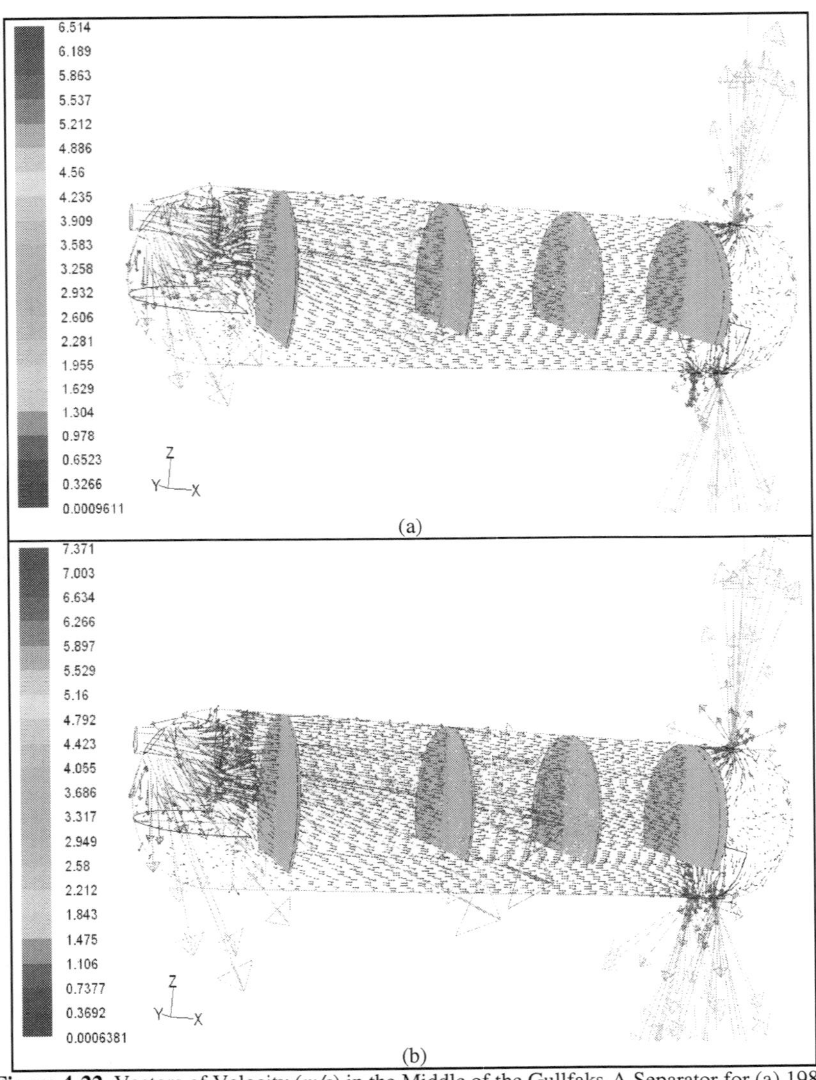

Figure 4-22. Vectors of Velocity (*m/s*) in the Middle of the Gullfaks-A Separator for (a) 1988, and (b) the Future Condition.

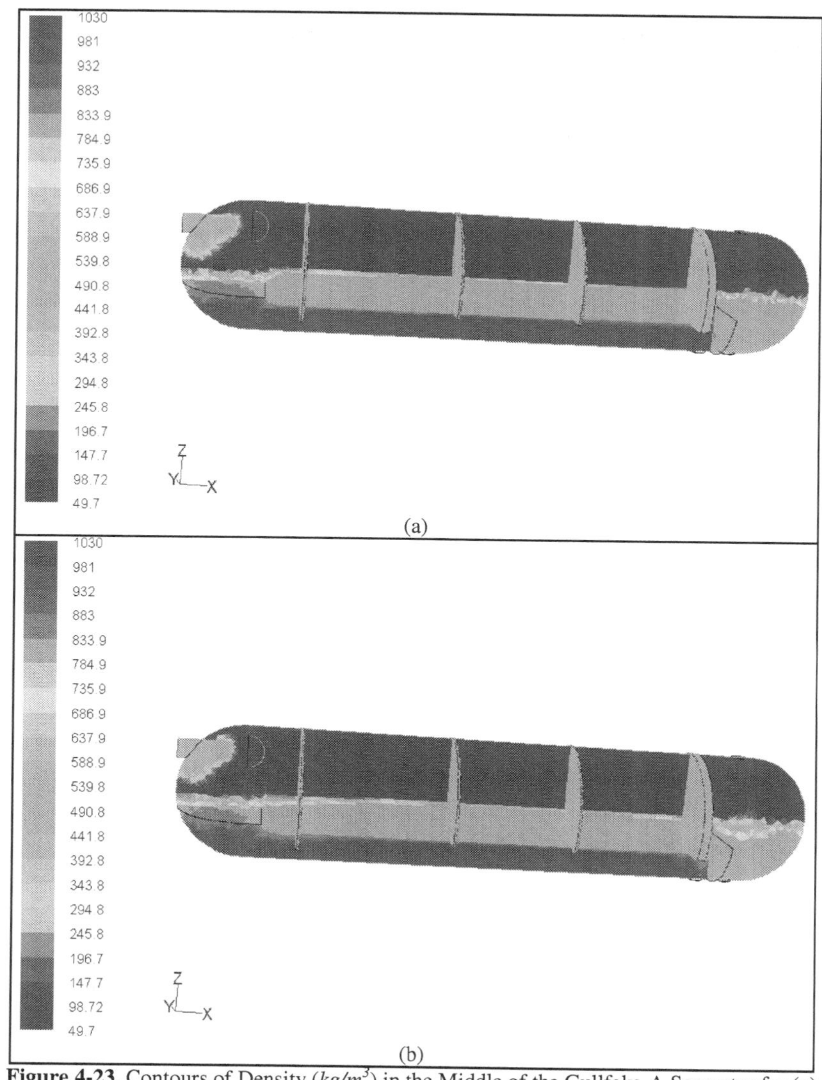

Figure 4-23. Contours of Density (*kg/m³*) in the Middle of the Gullfaks-A Separator for (a) 1988, and (b) the Future Condition.

4.2.4 *Separation Efficiencies*

The analysis of the oil and water droplets exiting at the separator outlets resulted in predicting a total separation efficiency of 98.0% at the 1988 production conditions. The result is based on mass distribution of injected oil and water droplets among the separator outlets. This mass distribution analysis indicated that 100% of oil droplets and 96.9% of water droplets were separated and came out through their corresponding outlets. Note, there were no droplets present in the gas phase outlet, hence all the injected droplets came out in either the oil outlet or the water outlet.

As expected from practical field experience and also shown in the density contours of Figure 4-23b, with an increase in the produced water flow-rate in the upcoming years, the separation efficiency for oil droplets was predicted to decrease to 1.3%. Again, there was no predicted carry over in the gas outlet, thus all the injected droplets came out with either the oil outlet or the water outlet. So, gas-liquid separation efficiency was still 100%. The very low separation efficiency for oil droplets indicates that water phase does not provide enough residence time for oil droplets to rise up and join oil phase, and almost all of oil droplets are carried by water phase to the water outlet. Although the separation efficiency was calculated to be 100% for water droplets, because of the difficulty in separating the oil droplets, the total separation efficiency has been reduced to 70.4%.

The result of droplet size distribution analysis on the selected surfaces of the separator is shown in Table 4-4. As Table 4-4 shows, compared with the initially defined droplet size distribution (Table 3-12), droplets have become smaller. The reason is that droplet breakup occurs when the injected droplets strike the deflector baffle. Therefore, the volume median diameter has decreased to 70% of its initial value for oil droplets, and to 67% of its initial value for water droplets for the 1988 production condition. With upcoming increase in the inlet water flow-rate, these values change to 54% for oil droplets and to 46% for water droplets. Table 4-4 also shows that droplet size distribution before and after gravity separation zone is almost the same. This implies that there should be no further breakup while droplets are traveling through the main part of the separator; hence, droplet size distribution remains essentially constant.

Table 4-4. The droplet size distribution in important zones of the Gullfaks-A separator.

		Discrete Phase Parameters	Before Gravity Separation Zone	After Gravity Separation Zone	Oil Outlet	Water Outlet
1988 Production Condition	Oil Droplets	d_{min} (μm)	26	26	26	———
		d_{max} (μm)	1848	1830	1209	———
		$\bar{d}$ (μm)	640	627	419	———
		n	3.18	3.20	3.80	———
	Water Droplets	d_{min} (μm)	85	85	93	90
		d_{max} (μm)	2424	2338	438	2315
		$\bar{d}$ (μm)	1077	1009	257	1008
		n	2.81	2.79	6.79	3.05
Future Production Condition	Oil Droplets	d_{min} (μm)	37	37	47	25
		d_{max} (μm)	1009	1009	219	855
		$\bar{d}$ (μm)	422	423	154	300
		n	4.42	4.43	6.55	3.60
	Water Droplets	d_{min} (μm)	34	34	76	34
		d_{max} (μm)	1596	1596	148	1593
		$\bar{d}$ (μm)	643	633	132	545
		n	4.11	3.55	6.60	3.67

The CFD simulation results show that for the 2000 injected droplets, the number of breakups was predicted to be 1590 for 1988 condition and 1543 for the future condition. Similar to the case of two-phase separators, free coalescence of droplets was not a common phenomenon in the three-phase separator. Droplet coalescence may happen at a very low rate of about 0.1%, without any noticeable trend.

4.2.5 *The Effect of Minor Modifications on the Gullfaks-A Separator*

The first modification that was tested was optimizing the position of the distribution baffles. To test this effect, all the baffles were removed from the model to see if any tangible effect on the separator performance might be observed. With this change, the separation efficiency was calculated as 99.5% for oil droplets and 96.8% for water droplets at the 1988 conditions, and 0.8% for oil droplets and 100% for water droplets at the future production conditions. Compared with the reported values for the separator equipped with the baffles, a negligible difference is seen. On the other hand, as concluded by Lu et al. (2007) in their recent CFD-based study, it can be expected that the baffles will improve the quality of liquid flow distribution in the vessel and increase the separation efficiency. Noting that the baffles were originally installed in the upper (vapor disengagement) section of the separator, it seemed that by expanding the baffles so that they cover the lower areas as well, some improvements might be achieved. In order to investigate this issue, baffles in the inefficient operating separator were expanded to cover the whole cross-sectional area. The resultant fluid flow profiles are presented in the supplementary package of the book, and the droplet size distributions for all the case studies discussed above are presented in Tables 4-5 and 4-6.

Table 4-5. The droplet size distribution in the important zones of the Gullfaks-A separator without the internals (flow-distributing baffles and the mist eliminator).

		Discrete Phase Parameters	Before Gravity Separation Zone	After Gravity Separation Zone	Oil Outlet	Water Outlet
1988 Production Condition	Oil Droplets	d_{min} (μm)	58	58	19	62
		d_{max} (μm)	1852	1852	1843	102
		$\bar{d}$ (μm)	702	702	435	89
		n	4.59	4.59	2.11	7.26
	Water Droplets	d_{min} (μm)	56	56	149	56
		d_{max} (μm)	2106	2106	495	2106
		$\bar{d}$ (μm)	1073	969	260	968
		n	2.95	2.98	4.34	3.01
Future Production Condition	Oil Droplets	d_{min} (μm)	34	24	81	14
		d_{max} (μm)	916	916	212	842
		$\bar{d}$ (μm)	430	434	179	274
		n	4.00	4.02	9.79	2.68
	Water Droplets	d_{min} (μm)	40	40	———	27
		d_{max} (μm)	1532	1532	———	1371
		$\bar{d}$ (μm)	653	653	———	475
		n	4.27	4.28	———	3.79

Table 4-6. The droplet size distribution in the important zones of the Gullfaks-A separator with expanded flow-distributing baffles in the future conditions.

	Discrete Phase Parameters	Before Gravity Separation Zone	After Gravity Separation Zone	Oil Outlet	Water Outlet
Oil Droplets	d_{min} (μm)	39	39	55	20
	d_{max} (μm)	1398	1398	191	1398
	$\bar{d}$ (μm)	492	493	172	346
	n	4.15	4.16	4.60	2.77
Water Droplets	d_{min} (μm)	34	34	———	34
	d_{max} (μm)	1557	1557	———	1475
	$\bar{d}$ (μm)	662	662	———	498
	n	3.79	3.79	———	3.31

The CFD simulation of droplet dispersions within the separator resulted in separation efficiencies of 0.5% and 100% for oil and water droplets, respectively. Therefore, it can be concluded that adjusting the arrangement of distributing baffles cannot really improve the separation efficiency in the Gullfaks-A separator. In fact, as emphasized by Lyons and Plisga (2005), although properly designed internals are generally helpful in reducing liquid carryover at design conditions, they cannot overcome separation problems in a basically inefficient design or in an undersized separator.

The other modification analyses were performed by changing the liquid levels. Although in all of the following liquid level analyses, the separation efficiency remained about 100% for water droplets, the separation efficiency did generally decrease for the oil droplets to be:

- 0.08% with an increase of 20 *cm* in the thickness of water layer,
- 0.03% with a decrease of 20 *cm* in the thickness of water layer, and
- 0.02% with an increase of 20 *cm* in thickness of both water and oil layers.

Thus, it may be concluded from the CFD simulations that minor modifications cannot change the separation inefficiencies in the Gullfaks-A separator.

4.2.6 *Redesign of the Gullfaks-A Separator*

The CFD simulation results showed that a major separation inefficiency would be encountered by the existing separator for the upcoming increase in the produced water flow-rate. The CFD simulations also showed that minor adjustments could not mitigate this separation inefficiency. Therefore, redesign of the separator is the only reasonable solution. To accomplish this redesign task, two approaches would be used. In the first approach, the separator was redesigned by one of the well-known classic methods. In the second strategy, the same classic design method was improved by considerations of appropriate retention times of separator at 1988 production conditions.

4.2.6.1 Classic Design Strategy

The algorithmic design method proposed by Monnery and Svrcek (1994) was used. The method uses industry standard settling or rising velocities to design the most economical separator. Unfortunately, the exact values of settling or rising velocities of droplets may be totally different from the industry accepted values. As noted in Chapter Two, the usual value for liquid droplet size is 150 μm and is used as a standard in the API design method (Walas, 1990; Hooper, 1997). However, assuming this droplet size, no realistic results could be obtained for the Gullfaks-A separator. That is, a separator with diameter of 14.4 m and length of 87 m would result for separating the mixture at 1988 production conditions, and a separator with dimensions of 12.4 m $\times$ 75 m would be proposed for the future production conditions. The calculated separator dimensions imply that, through classic design strategy, there is no feasible design for separating the mixture with the Gullfaks-A production conditions. On the other hand, both CFD simulation results and the oilfield experience showed that the original separator has operated in an acceptable way for the 1988 conditions, and the separation inefficiencies have been associated with the increase in the produced water flow-rate at the future production conditions. This would indicate that a realistic design could be made by combining the classic method with new separation velocities developed from CFD simulations.

4.2.6.2 Combining Improved Separation Velocities

As noted by Svrcek and Monnery (1993), although the basic equations used for separator sizing are widely known; subjectivity exists during the selection of the parameters used in these equations. The most controversial parameters are settling (or rising) velocities of droplets in liquid-liquid separation. These parameters are to be tuned via CFD simulation using the 1988 production conditions at which the separation efficiencies have been acceptable.

- Estimation of settling velocity of oil droplets in the gas phase:
  Two industrial methods, i.e. Gas Processors Suppliers' Association (GPSA) method and York demister method, and one theoretical method were evaluated. Using these methods, the terminal settling velocity of oil droplets was calculated to be 0.460 *m/s*, 0.328 *m/s*, and 0.216 *m/s*, respectively. Thus, the corresponding minimum length required for the vapor-liquid separation in the original separator was calculated as 0.51 *m*, 0.71 *m*, and 1.08 *m*, respectively. The CFD simulations indicated that around 10 *cm* after the momentum breaker, almost all the liquid droplets have settled down, which implies that the calculated values are somewhat conservative. However, while sizing the Gullfaks-A separator, the classic procedure did indicate that the liquid-liquid separation is controlling. Therefore, the focus of the current phase of this study was placed on the proper estimation of oil-water separation velocities.

- Estimation of the rising velocity of oil droplets out of the water phase:
  In classic methods, the flow of rising oil droplets in the water phase is assumed to be laminar and the rising velocity can be estimated using Stokes' law. By using Stokes' law, a velocity of 0.00566 *m/s* was calculated for the industry accepted droplet size of 150 μm. This value is higher than the value implied by CFD simulation results as follows. The separation efficiency of 100% for oil droplets indicates an appropriate separator design/size. The provided residence time of this original separator was 165.6 *s* for water phase with a water layer thickness of 0.625 *m*, hence the rising velocity of oil droplets should be 0.00377 *m/s* as was assumed while redesigning the separator.

- Estimation of the settling velocity of water droplets out of the oil phase:
  The flow of settling water droplets in the oil phase is assumed to be laminar, hence Stokes' law is used in the classic methods. A velocity of 0.000464 *m/s* was calculated assuming a liquid droplet size of 150 μm. Compared with the CFD simulation results, this value is too conservative and leads to an oversized separator. The installed separator with dimensions of 3.33 $m \times$ 16.30 m has been working efficiently for the designed 1988 production conditions. From a CFD point of view, the predicted separation efficiency of 96.9% for water droplets indicates that the design has been successful in having the water droplets separated from the oil phase. Therefore, the provided residence time of 73.4 *s* for oil phase was reasonable. Accounting for the oil layer thickness of 1.039 *m*, the settling velocity of water droplets should be 0.01416 *m/s*. In order to further improve the water separation efficiency, a value of 0.0134 *m/s* (95% of 0.01416 *m/s*) was assumed while redesigning the separator.

Instead of the estimated values calculated from Stokes' law, new "separation velocities" were used in the classic design method proposed by Monnery and Svrcek (1994). In order to specify a useable separator, the separator was redesigned for the maximum flow-rates of produced gas, oil, and water. Figure 4-24 provides the sizing specifications of the redesigned Gullfaks-A separator. As expected, a larger separator is required. To keep the design in line with the original separator, four extra baffles with gaps of 4.0 *m* between them were also added to the new separator design. Figure 4-25a shows the vessel in the Gambit environment after assigning the required mesh, and Figure 4-25b represents the separator in the Fluent environment with all the installed internals. The global quality of the produced mesh in terms of number of cells, maximum cell squish, cell skewness, and maximum aspect ratio are presented in Table 4-7. Table 4-7 indicates that only a negligible fraction of cells (around 0.0044%) was of poor quality, hence the mesh is correct.

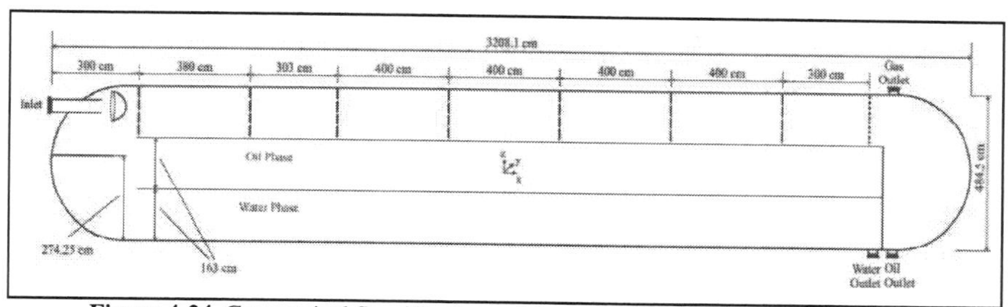

Figure 4-24. Geometrical Specifications of the Redesigned Gullfaks-A Separator.

Table 4-7. Quality of the mesh produced in Gambit environment for the redesigned Gullfaks-A separator.

Number of Cells	Maximum Squish	Maximum Skewness	Maximum Aspect Ratio
1155813	0.961646	0.999632	128.621

Skewness of the Produced Mesh					
Skewness Range	0-0.20	0.20-0.40	0.40-0.60	0.60-0.80	0.80-1.0
Density of Cells	76.4772%	17.8354%	4.4223%	1.2607%	0.0044%

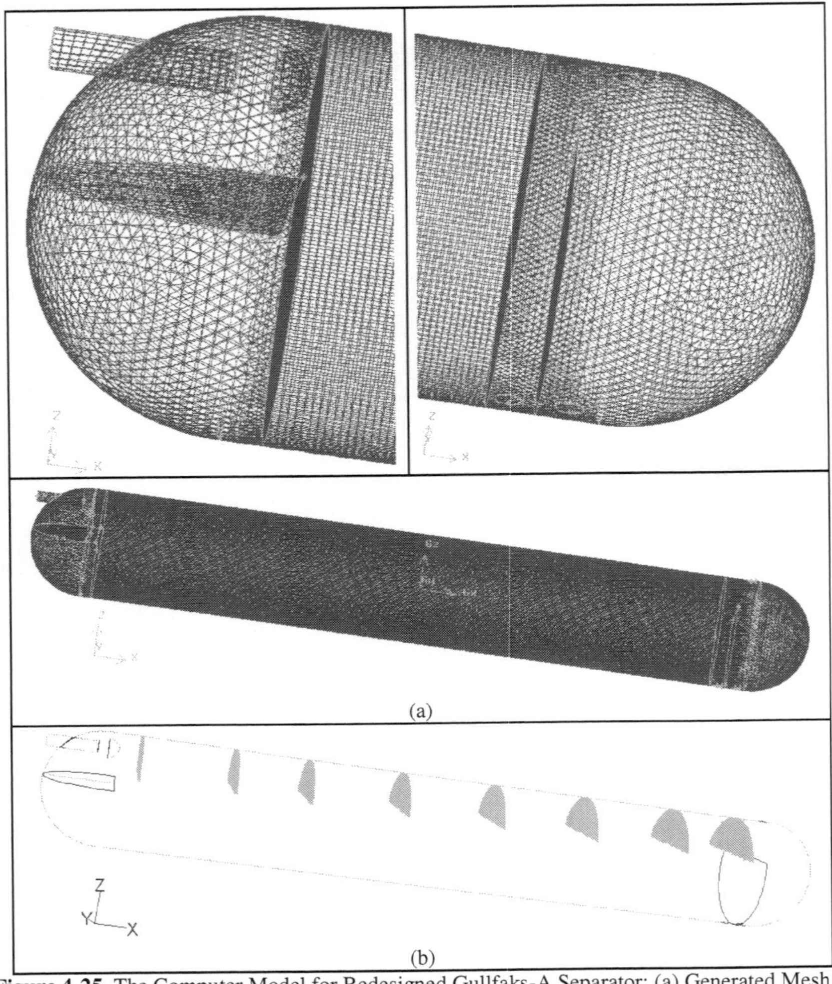

Figure 4-25. The Computer Model for Redesigned Gullfaks-A Separator; (a) Generated Mesh in Gambit Environment, and (b) with All the Internals in Fluent Environment.

Having set all the CFD parameters for the redesigned separator, some 5100 iterations are required for continuous-phase solution convergence. Each iteration takes about 31 *s* and a PC run-time of around 44 *h* is hence required per solution of the continuous phases. A further PC run-time of around 4 *h* is also required for simulation of interactions between the dispersed droplets and continuous phases. The fluid flow profiles inside the redesigned separator are presented in Figures 4-26 to 4-28. These figures provide the pressure, velocity and density profiles for the fluid flows on the central xz-plane of the separator. As was expected, the macroscopic features of the redesigned separator are very similar to those of the original separator. Again the pressure drops predicted for the baffles and demister are low and hence reasonable, and the velocity vectors and density contours are realistic.

The modifications made as a result of the CFD simulation of this study did enhance the separator performance. The redesigned separator dealt satisfactorily with 1988 production conditions, in that the total separation efficiency was as high as 99.1% (a bit higher than the original separator efficiency) with its components of 100% and 98.7% as separation efficiencies for oil and water droplets, respectively. For the projected future production conditions, the redesigned separator had a total separation efficiency of 99.7% with its components of 93.4% and 100% as separation efficiencies for oil and water droplets, respectively. Note, although some 6.6wt% of the oil droplets were predicted to exit in the water outlet for future production case, this would not decrease the separated water purity dramatically, in that the water outlet composition would be 99.7% water and only 0.3% oil. As was the case with the original separator, there would be no droplet carry-over in the gas phase outlet, i.e. all the injected droplets exited in either the oil outlet or the water outlet. Droplet size distribution is given for the selected surfaces of the separator in Table 4-8. Table 4-8 indicates that compared with the initially defined droplet size distribution (Table 3-12) the volume median diameter has decreased to become 87% of its initial value for oil droplets and to 75% of its initial value for water droplets. At the new project operating conditions the values changed to 63% for oil droplets and to 73% for water droplets.

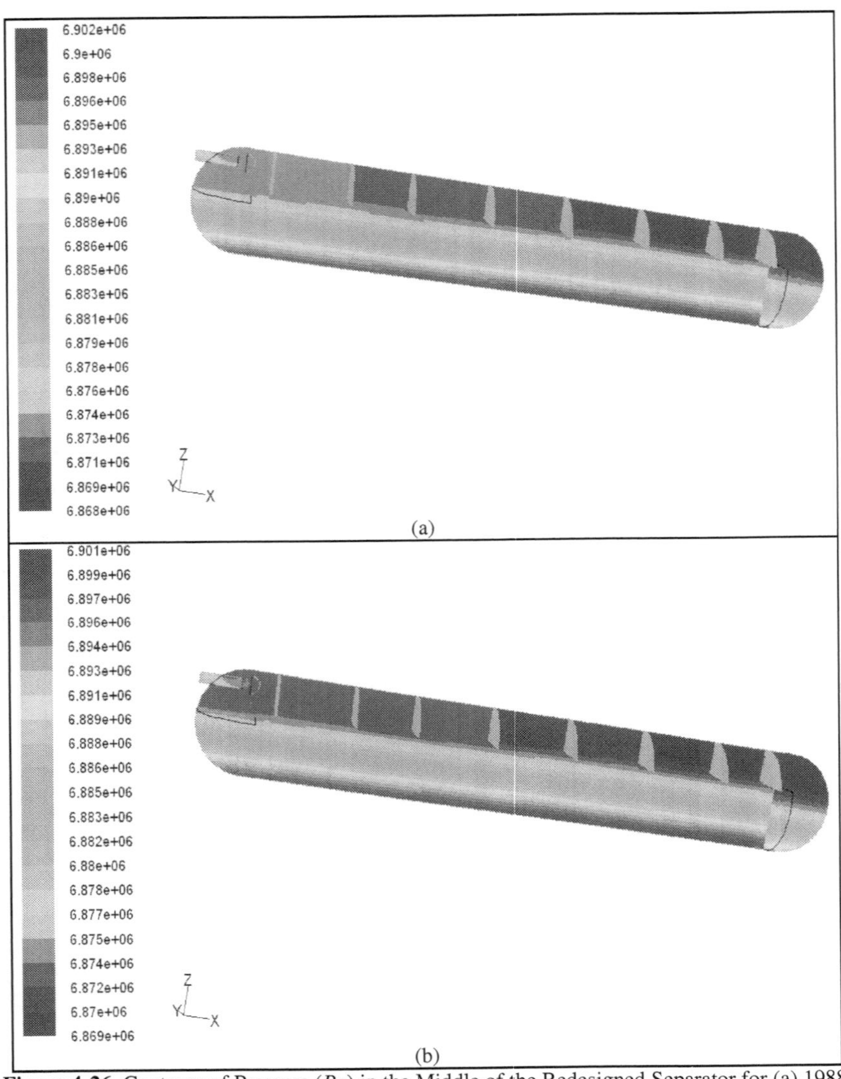

Figure 4-26. Contours of Pressure (*Pa*) in the Middle of the Redesigned Separator for (a) 1988, and (b) the Future Conditions.

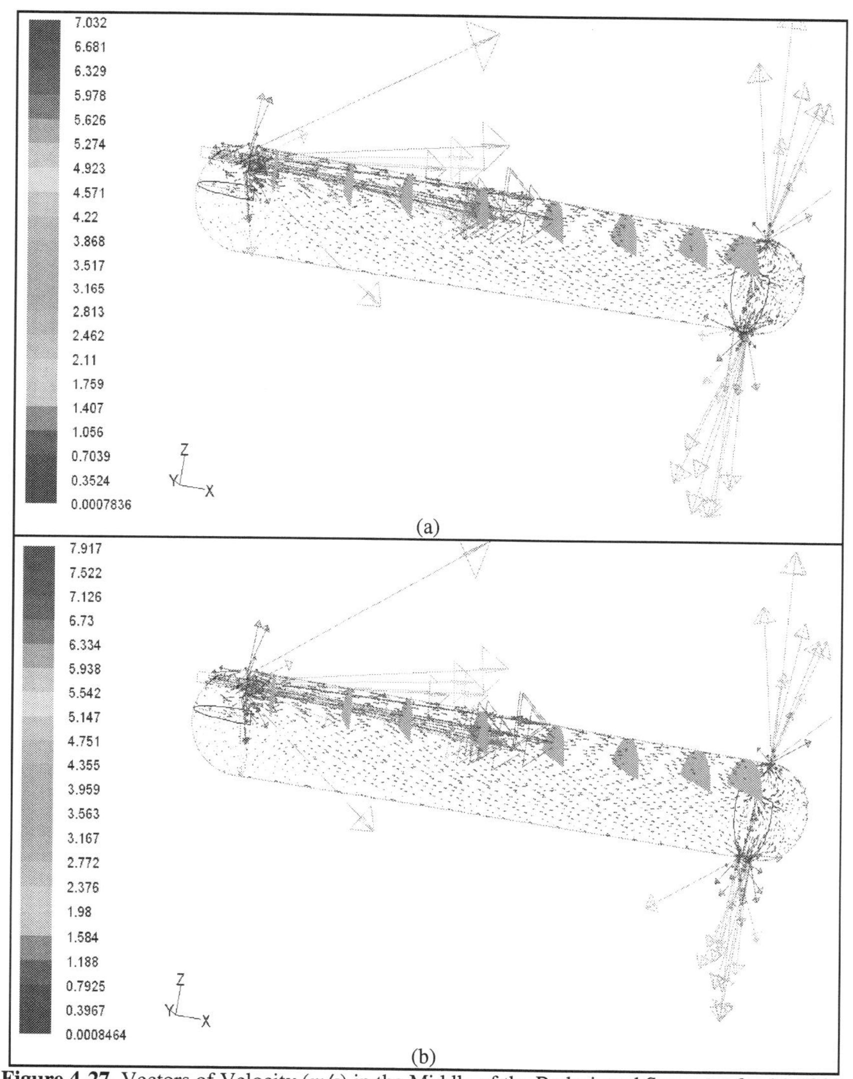

Figure 4-27. Vectors of Velocity (*m/s*) in the Middle of the Redesigned Separator for (a) 1988, and (b) the Future Conditions.

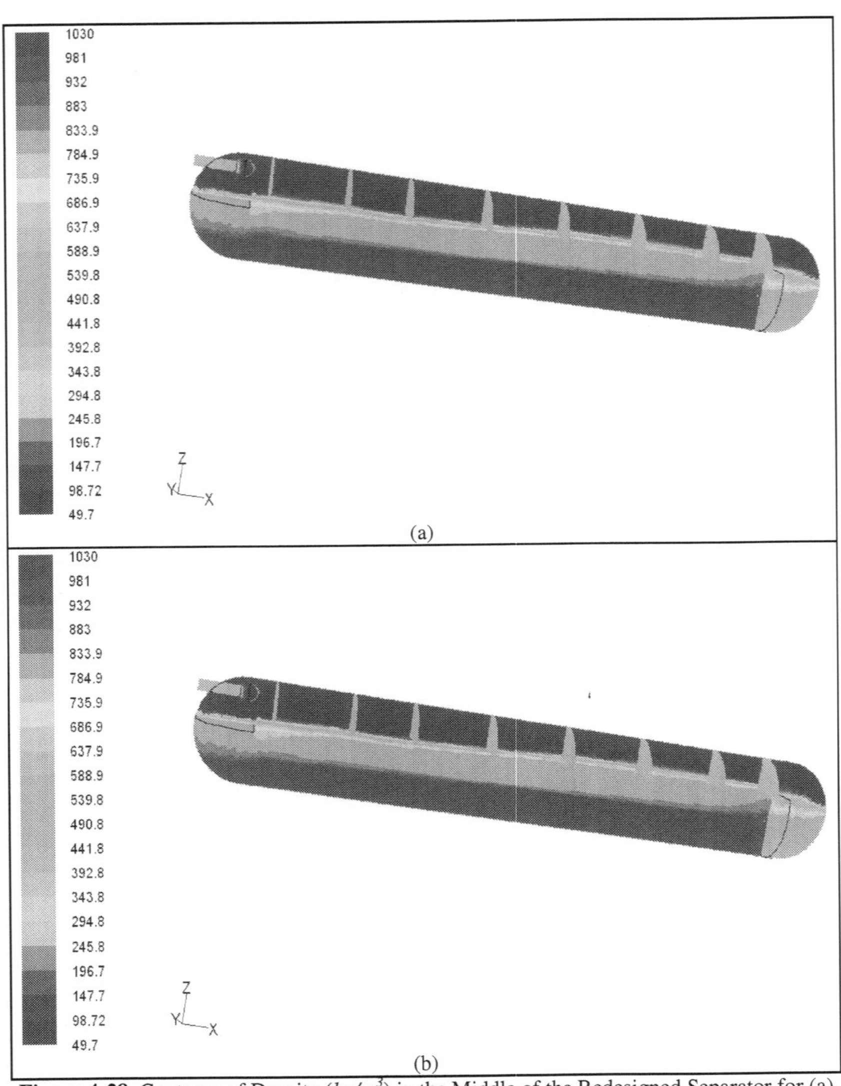

Figure 4-28. Contours of Density (kg/m^3) in the Middle of the Redesigned Separator for (a) 1988, and (b) the Future Conditions.

Table 4-8. The droplet size distribution in the important zones of the redesigned Gullfaks-A separator.

		Discrete Phase Parameters	Before Gravity Separation Zone	After Gravity Separation Zone	Oil Outlet	Water Outlet
1988 Production Condition	Oil Droplets	d_{min} (μm)	17	17	17	———
		d_{max} (μm)	2058	2058	1755	———
		$\bar{d}$ (μm)	791	799	683	———
		n	2.52	2.18	2.19	———
	Water Droplets	d_{min} (μm)	29	16	29	33
		d_{max} (μm)	2509	1615	476	496
		$\bar{d}$ (μm)	1218	390	403	223
		n	2.95	2.27	1.98	2.04
Future Production Condition	Oil Droplets	d_{min} (μm)	17	25	39	25
		d_{max} (μm)	1835	1846	1504	352
		$\bar{d}$ (μm)	490	683	664	270
		n	3.94	2.47	5.62	2.28
	Water Droplets	d_{min} (μm)	40	35	———	35
		d_{max} (μm)	3007	3007	———	1240
		$\bar{d}$ (μm)	1004	1074	———	480
		n	2.54	3.28	———	4.07

While tracking the 2000 injected droplets, the number of breakups predicted in the CFD simulation was some 1363 for 1988 production conditions and 1568 for the future production conditions.

Similar to the case of the original separator, the effect of installed distribution baffles was simulated. Again, all the baffles were removed from the model and this was assumed to be a base case. The resultant separation efficiency was calculated to be 100% for oil droplets and 99.3% for water droplets at the 1988 production conditions. For the future production conditions, a separation efficiency of 97.7% for oil droplets and of 100% for water droplets was calculated. When compared with the corresponding values reported for the separator equipped with the baffles, some small differences in favor of eliminating the baffles results.

The second phase of the internals simulation again consisted of expanding the baffles so that they cover all the cross-sectional area of the vessel. The results of simulation in terms of the pressure, velocity and density profiles are presented in the supplementary package of the book. The results of the CFD modeling predicted efficiencies of 100% and 99.7% for the oil and water droplets at the 1988 production conditions, respectively. Thus, the separation efficiency of water droplets has shown a minor improvement with the baffles expanded in the separator. However, a noticeable decrease in the oil separation efficiency (from 93.4% to 75.2%) was predicted for the separator operating at the future production conditions. For this case, the predicted separation efficiency was still 100% for the water droplets.

The results generated for the original separator and its modified versions would point to the fact that distribution baffles have a minor effect on the separation efficiency. They are generally helpful in improving the quality of flow distribution, however, installing distribution baffles cannot overcome poor designs. Moreover, there may be some rare cases with decreased separation efficiency after installing the baffles.[10]

4.2.6.3 On Substitution of Stokes' Law by More Complicated Models

Using Stokes' law in the classic design method led to an oversized separator for Gullfaks-A. In fact, Stokes' law assumes a single solid particle falling or rising in a stationary fluid and calculates the particle terminal velocity. This, of course, is not the case in an actual multiphase separator, in which a population of fluid droplets (of different size, position and velocities) is to be separated from a moving fluid flow. So, the complicated aspects of the described process make it impossible to come up with a mathematical model to precisely describe the process. However, some modified relations have been proposed to mitigate the Stokes' law restrictions. The relations provided by Ishii and Zuber (1979) are useful in that the authors have provided empirical relations for estimation of drag coefficients and relative terminal velocities for dispersed droplets. The relations are rather complicated, as expected, and only the pertinent equations are presented here. Equation 4-1 has been proposed by Ishii and Zuber (1979) for

[10] Please also refer to Hansen et al. (1993) for a practical example.

calculation of the relative terminal velocity in the "undistorted droplet regime", which satisfies the Gullfaks-A separation conditions:

$$U_d = \frac{21.6\mu_c}{\rho_c d_p}\frac{\mu_c}{\mu_m}(1-\beta)\frac{\varphi^{4/3}(1+\varphi)}{1+\varphi\left[\frac{\mu_c}{\mu_m}\sqrt{1-\beta}\right]^{6/7}} \quad \textbf{(4-1)}$$

where U_d is relative velocity of droplets in *m/s*, d_p is diameter of droplets in *m*, μ_m is mixture viscosity in *Pa.s* as defined by Equation 4-2, β is volume fraction of dispersed phase, and φ is a function of d_p as given by Equation 4-3.

$$\frac{\mu_c}{\mu_m} = (1-\beta)^{2.5(\mu_d+0.4\mu_c)/(\mu_d+\mu_c)} \quad \textbf{(4-2)}$$

$$\varphi = 0.55\left[\left(1+\frac{0.01\rho_c|\rho_d-\rho_c|d_p^{\ 3}g}{\mu_c^{\ 2}}\right)^{4/7}-1\right]^{0.75} \quad \textbf{(4-3)}$$

Using Equations 4-1 to 4-3 and assuming a volume fraction of 5% for the dispersed phase and the usual droplet size of 150 μm, the separation velocities are calculated to be 0.00452 *m/s* and 0.000416 *m/s* for oil and water droplets, respectively. It is interesting that the calculated separation velocities partially agree with the Stokes' law results, with a difference of 25.2% for oil droplet velocity and 11.5% for water droplet velocity. In order for the model proposed by Ishii and Zuber (1979) to predict the more realistic separation velocities, the oil and water droplet diameters should be assumed to be 135 μm and 876 μm, respectively. The corresponding oil and water droplet diameters for proper velocity estimation by Stokes' law are 122 μm and 806 μm, respectively. Again, there are small differences of about 9% between the corresponding results for the Stokes' law and Ishii and Zuber equations. Therefore, it would seem that having assumed proper/efficient droplet sizes, the separation velocities can be estimated by simpler models such as Stokes' law. This issue will be discussed in more detail in Chapter Five.

4.2.7 *Sensitivity Analyses*

In this subsection, sensitivity of the results with respect to the most important parameters used in the CFD simulations is discussed. The first issue of interest is the repeatability of the CFD predicted separation efficiencies. For this purpose, the original Gullfaks-A separator CFD simulation at the two production conditions was used and the injected droplets were tracked for four times. A statistical analysis of the produced results indicated that the standard deviation of results was generally less than 1% for the average values in all cases. Therefore, the number of performing DPM model of Fluent was reduced to one in the modification case studies. The other important CFD parameters for sensitivity analyses are as follows:

1- Assigned Grid Size

Quality of the produced mesh for Gullfaks-A separator was verified in Chapter Three. As discussed, the grid system was composed of 884805 cells, corresponding to an average interval size of 0.037 *m*. However, in order to further verify the quality of applied grid system, a relatively coarse grid system, composed of 500130 cells, with an average interval size of 0.065 *m* was used to see if the CFD results would change significantly. The tendency with denser grids is to reveal more details of the fluid flow regimes, and if the obtained solution remains unchanged in its important features, it can be concluded that the assumed grid system is sufficiently fine (Sharratt, 1990).

2- Droplet Size Distribution

To analyze the sensitivity of the CFD results to the defined droplet size distribution, the maximum droplet size was alternatively estimated by the Karabelas (1978) correlation instead of Hesketh et al. (1987) approach, and the mean droplet size was changed using Equation 3-24.

3- Surface Tensions

The surface tension values were estimated using the methods discussed in Chapter Three. However, for sensitivity analysis purposes, the oil and oil-water surface tensions were set to 0.038 *N/m*, the maximum value given by Streeter and Wylie (1985), and 0.03 *N/m*, based on Antonoff's rule (1907), respectively.

After implementing these changes in the CFD model or its parameters, the fluid flow profiles were obtained for each of the case studies and were compared with those of the original case study. The CFD results were also compared from a separation efficiency point of view. These comparisons indicated that both the macroscopic and microscopic features of the CFD solution for Gullfaks-A separator were not so sensitive to the applied changes. For example, Table 4-9 provides the separation efficiencies of the studied cases as well as the original Gullfaks-A values. The data of Table 4-9 would indicate that there are some negligible differences between the "sensitivity analysis" case studies and the original separator values. Therefore, it can be concluded that the original grid system is sufficiently fine and that some uncertainties in the defining droplet size distribution or estimated surface tensions have little effect on the CFD simulation results of multiphase separators.

Table 4-9. The results of sensitivity analyses on the developed CFD model for Gullfaks-A separator.

		Separation Efficiencies			
		1988 Condition		Future Condition	
		Oil	Water	Oil	Water
Sensitivity Analysis Parameter	Grid Size	98.0%	99.8%	0.0%	100%
	Maximum Particle Size	100%	99.6%	2.1%	99.8%
	Surface Tension	99.9%	96.3%	0.1%	100%
Original Values		100%	96.9%	1.3%	99.9%

4.3 Summary

In Chapter Four, the CFD simulation results of selected pilot-plant scale two-phase separators and one large scale three-phase separator have been presented. Two simulation approaches, DPM and VOF-DPM, were used in the simulation of the two-phase separators. The DPM approach was useful in predicting the incipient velocities, but the resultant separation efficiency plots were not correct. In fact, some unacceptable results were obtained for several case studies when using the DPM only model as proposed by Newton et al. (2007). These poor results were caused by the model assumption in that the liquid phase is ignored. Thus, the gas-liquid interactions and, particularly, the dynamic interaction between liquid droplets and continuous liquid phase cannot be taken into account. The more complex VOF-DPM model does include these liquid phase interactions. From a practical perspective, the obtained separation efficiency data and diagrams were reasonable and the poor results of the DPM model were completely eliminated. Moreover, there was excellent agreement between simulated phase separation results, such as separator efficiencies etc., and the experimental data and observations in most of the case studies. Droplet coalescence and breakup were also modeled through the CFD simulations. The results indicated that while droplet coalescence occurred rarely, droplet breakup was a common phenomenon particularly at higher velocities. The study of droplet size distribution in the gas outlet for the selected separators showed that mist eliminators may operate more efficiently in the horizontal separators than in the vertical separators.

Based on the obtained results for the CFD simulation of two-phase separators, only VOF-DPM approach was used in the CFD simulation of the three-phase separator. Compared to the original study of Hansen et al. (1993), the developed model did provide high-quality details of fluid flow profiles, leading to a very realistic overall picture of phase separation in all zones of the separator. This realistic CFD simulation of the three-phase separator performance did provide an understanding of both the microscopic and macroscopic features of the three-phase separation phenomenon. The CFD simulations did show that droplet breakage was common with an average rate of 76%, when dispersed droplets came into contact with the deflector baffle. Because of droplet breakup, the volume median diameter of droplets decreased to about 67% of the initial value. However, the droplet size distribution remained almost the same while the droplets were traveling through the gravity separation zone of the separator.

In line with the oilfield experience, the CFD simulations showed that serious separation inefficiencies may be encountered with the projected increase in the flow-rate of produced water in the upcoming years. To overcome these inefficiencies, first, minor modifications such as adjusting baffle positions and liquid levels were evaluated. However, these minor modifications could not resolve the essential separation inefficiencies in the Gullfaks-A separator, and the separator was redesigned using a modified version of the classical design method of Monnery and Svrcek (1994) that involved improved separation velocities. It was also realized that the popular classic methods would be too conservative and specify an extremely oversized separator for Gullfaks-A. The redesigned separator operated satisfactorily at both the 1988 and the projected production conditions and did confirm that a realistic optimum separator can be specified using the classical method of Monnery and Svrcek (1994) when the new developed separation velocities are incorporated in the design procedure. Furthermore, it was demonstrated that simple models such as Stokes' law may still be used for estimation of realistic separation velocities, but suitable/efficient droplet sizes should be used for the oil and water phases.

Finally, sensitivity analysis was performed on the developed CFD model for Gullfaks-A separator. The results reconfirmed the quality of the grid system and robustness of the DPM outputs with respect to solution repetitions. The sensitivity analysis also showed that small (<10%) uncertainties in droplet size distribution or in estimating surface tensions will have negligible effect on the CFD simulation results of multiphase separators.

Chapter Five: Improved Design Criteria

CFD simulations of four pilot-plant scale two-phase separators and an industrial scale three-phase horizontal separator were developed. In order to successfully simulate these separators, an efficient combination of two multiphase simulation models of Fluent, VOF and DPM, with appropriate model assumptions and settings was used. The obtained results were in excellent agreement with the empirical observation and pilot plant data. In the case of three-phase separator, the developed model did provide useful fluid flow profiles and a more realistic picture of phase separation quality when compared with the original study of Hansen et al. (1993). Hence, the goal of the present chapter is to keep the implemented model assumptions/settings and provide updated phase separator design criteria. These criteria can then be combined with the algorithmic design method of Monnery and Svrcek (1994) to specify the realistic optimum separator design.

5.1 Fluid Systems

The following oilfield separator data ranging from light oil conditions to heavy oil conditions were used to obtain the necessary physical properties required in the CFD simulations:

- Oilfield-1: Black Oil Reservoir Case Study (Grødal and Realff, 1999)
- Oilfield-2: Gullfaks A (Hansen et al., 1993)
- Oilfield-3: A Diluted Bitumen Treatment (Lu et al., 2007)
- Oilfield-4: Kuito Field Offshore Cabinda (Lee et al., 2004)
- Oilfield-5: Beta Offshore California (Visser, 1989)

The physical properties used in CFD simulations are presented in Table 5-1. Except for Oilfield-5, all the physical properties necessary for CFD simulations have been provided by original papers. In the case of Oilfield-5, the data for water and vapor phases are missing from original paper (Visser, 1989), and these data had to be estimated. Since the water phase density does not differ significantly from one oilfield to another, the value from Oilfield-4 was used in the simulations. For estimation of the water phase viscosity, the approach recommended by McCain (1973) was used. This approach estimates the water phase viscosity using the oilfield temperature and pressure. For estimation of vapor phase properties, the natural gas composition of Vetter et al. (1987) for a rather similar heavy oilfield (22.20°*API*) was used in the HYSYS 3.2 simulator. The recommended Peng-Robinson (PR) equation of state was used as the thermodynamic model for this gas phase hydrocarbon system.

As noted in Chapter Four, separation efficiencies are not that sensitive to the assumed maximum droplet size. Therefore, a maximum droplet size of 2267 μm and 4000 μm, as estimated for Gullfaks-A oilfield at the 1988 production condition, were assumed for the oil and water droplets, respectively. Then, the required particle size distribution parameters were set using the strategy outlined in Chapter Three. The results are shown in Table 5-2.

Table 5-1. Physical properties for fluids of different oilfields at their corresponding separator conditions.

		Oilfield-1	Oilfield-2	Oilfield-3	Oilfield-4	Oilfield-5
Pressure (*kPa*)		2000	6870	1280	690	345
Temperature (°*C*)		80	55.4	135	50	60
Gas	Density (kg/m^3)	17.46	49.7	9.50	5.70	2.44
	Viscosity (*Pa.s*)	1.07×10^{-5}	1.30×10^{-5}	1.38×10^{-5}	1.20×10^{-5}	1.26×10^{-5}
Oil	Density (kg/m^3)	767.7	831.5	874	907	960
	Viscosity (*Pa.s*)	0.73×10^{-3}	5.25×10^{-3}	6.90×10^{-3}	42.0×10^{-3}	100×10^{-3}
Water	Density (kg/m^3)	974.6	1030	931	1026	1026
	Viscosity (*Pa.s*)	3.70×10^{-4}	4.30×10^{-4}	2.0×10^{-4}	9.0×10^{-4}	5.2×10^{-4}

Table 5-2. The discrete phase parameters used in CFD simulations of phase separation.

Discrete Phase Parameters	Oil Droplets	Water Droplets
Minimum Diameter (μm)	100	100
Maximum Diameter (μm)	2267	4000
Mean Diameter (μm)	907	1600
Total Mass Flow-rate (kg/s)	6.5×10^{-4}	4.4×10^{-3}
Number of Tracked Particles	1000	1000
Spread Parameter	2.6	2.6

5.2 Phase Separation Components

In order to efficiently model the phase separation process, two independent sets of CFD simulations were performed; one for vapor-liquid separation and the other for liquid-liquid separation. For this purpose, two two-phase models shown in Figure 5-1 were used. These useful models were developed by verification of the results for various two-phase systems when compared to those of industrial scale separators. These investigations confirmed that although the mesh generation stages in the Gambit environment and the setting of CFD parameters in the Fluent environment would be more straightforward for these models, the results produced for phase separations are the same as those for large-scale separators.

The other reason for using these models is connected with their horizontal layout. By using the horizontal layout, the results can be directly used in designing horizontal separators, and the subjectivity of applying the relationships obtained from the vertical orientation force balance to the horizontal orientation is not encountered. Before reviewing the results, two additional comments are worth noting:

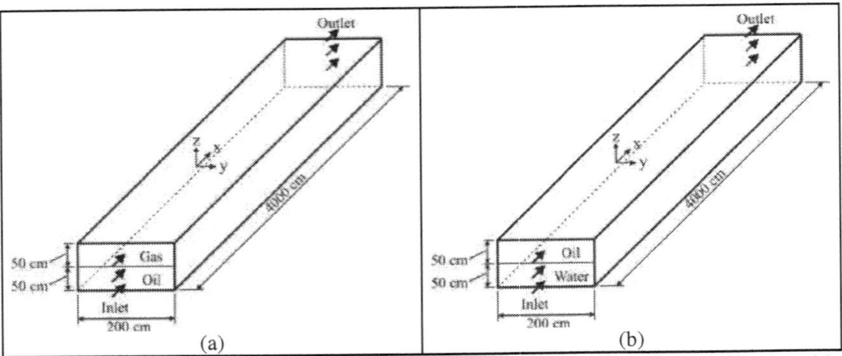

Figure 5-1. The Two-Phase Models Developed for CFD Simulation of Phase Separation; (a) Vapor-Liquid Separation, and (b) Liquid-Liquid Separation.

Comment 1: In the developed models, for the first time, a distribution of droplets is being tracked. Therefore, in order to evaluate the separation velocities, a mass averaged time taken by droplets to separate out of continuous phases was used. So, the reported values are the "efficient" separation velocities predicted for various oilfield systems. Having predicted the efficient separation velocities for a specific system, the corresponding efficient droplet size can then be calculated using available relationships.

Comment 2: In classic separator design methods, it is assumed that when oil or water droplets come in contact with vapor-liquid or liquid-liquid interfaces, they immediately penetrate the interface and join their own phases. Therefore, assuming an extra residence time for droplet penetration through interfaces has been ignored in the classic design methods. However, the CFD simulation results indicate that depending on the involved fluid properties, a residence time of up to 100 *s* may be necessary for droplets to pass through the interfaces.

5.2.1 *Vapor-Liquid Separation*

The CFD simulation results of vapor-liquid separation for various oilfields have been presented in Table 5-3. For calculation of the efficient droplet diameter, the well-known equation, Equation 5-1, was used (Green and Perry, 2008):

$$V_{sep} = \sqrt{\frac{4\times10^{-6} g d_{eff} (\rho_L - \rho_V)}{3C_D \rho_V}} \quad \textbf{(5-1)}$$

where V_{sep} is separation velocity in *m/s*, g is gravity acceleration in *m/s*^{2}, d_{eff} is efficient diameter of droplet distribution in μm, ρ_L and ρ_V are liquid and vapor densities (respectively) in *kg/m*^{3}, and C_D is drag coefficient which can be calculated based on the Gas Processors Suppliers' Association (GPSA) approach (Monnery and Svrcek, 2000) via Equations 5-2 and 5-3:

$$C_D = \frac{5.0074}{\ln(x)} + \frac{40.927}{\sqrt{x}} + \frac{44.07}{x} \quad \textbf{(5-2)}$$

$$x = \frac{3.35\times10^{-9}\rho_V(\rho_L-\rho_V)d_{eff}^{\ 3}}{\mu_V^{\ 2}} \qquad \textbf{(5-3)}$$

where μ_V is viscosity of vapor phase in *Pa.s*.

Table 5-3 presents the residence times required for oil droplets to pass through the vapor-liquid interface. The values vary from 4 *s* to 58 *s* almost linearly with respect to vapor phase density which varies from 2.44 kg/m^3 to 49.7 kg/m^3. For water droplets, the interface residence times decrease to about 4.5 *s* with a reported maximum value of 11.4 *s*. Also, the settling velocities of water droplets are higher than those of oil droplets in all cases. Therefore, in the vapor-liquid compartment of a three-phase separator, the efficient separation of oil droplets guaranties satisfactory separation of the water droplets.

Figure 5-2 shows the variations of calculated efficient oil droplet size versus gas phase density. The CFD simulation results shown in Figure 5-2 can be fit with an exponential function, Equation 5-4:

$$d_{eff,Oil-Gas} = 206.06 - 189.42\exp(-0.08635\rho_V) \qquad \textbf{(5-4)}$$

Table 5-3. The vapor-liquid separation characteristics for different oilfields.

		Oilfield-1	Oilfield-2	Oilfield-3	Oilfield-4	Oilfield-5
Oil Droplets	V_{sep} (*m/s*)	0.7215	0.4967	0.8623	0.9944	1.0869
	$t_{interface}$ (*s*)	12.4	58.0	30.0	14.5	4.0
	d_{eff} (μm)	164.7	203.6	119.3	94.6	51.0
Water Droplets	V_{sep} (*m/s*)	0.9691	0.7055	1.0620	1.1905	1.2644
	$t_{interface}$ (*s*)	4.0	5.9	5.5	11.4	2.0
	d_{eff} (μm)	223.6	311.6	163.5	116.3	62.7

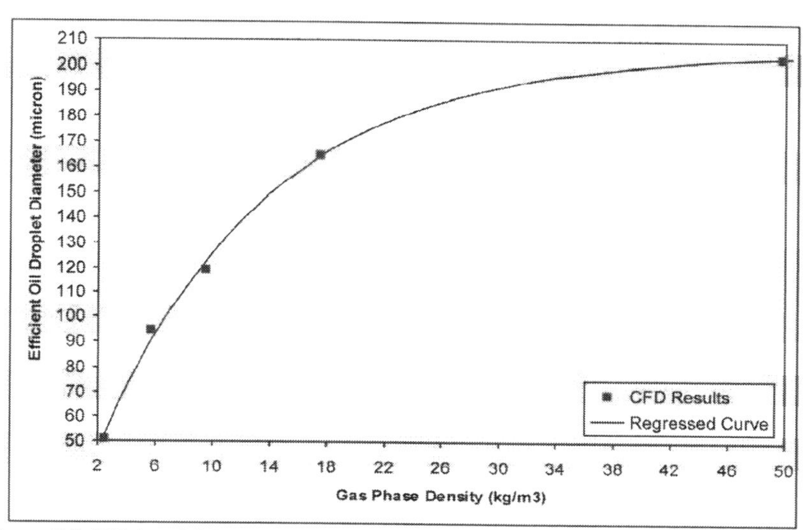

Figure 5-2. Efficient Oil Droplet Diameter for Estimation of Settling Velocity in Vapor Phase versus Vapor Density.

Similarly, Figure 5-3 shows the variations of calculated efficient water droplet size versus gas phase density, and Equation 5-5 presents the result of an exponential function fit for d_{eff} - ρ_V data:

$$d_{eff,Water-Gas} = 322.14 - 302.66\exp(-0.06619\rho_V) \qquad \textbf{(5-5)}$$

As noted, only oil-gas separation needs to be used for separator design. Therefore, Equations 5-1 to 5-4 can be used for estimation of vapor-liquid separation velocity. Furthermore, an additional vapor residence time proportional to the vapor density (or about 60 *s* for a conservative design), should also be used in the separator design.

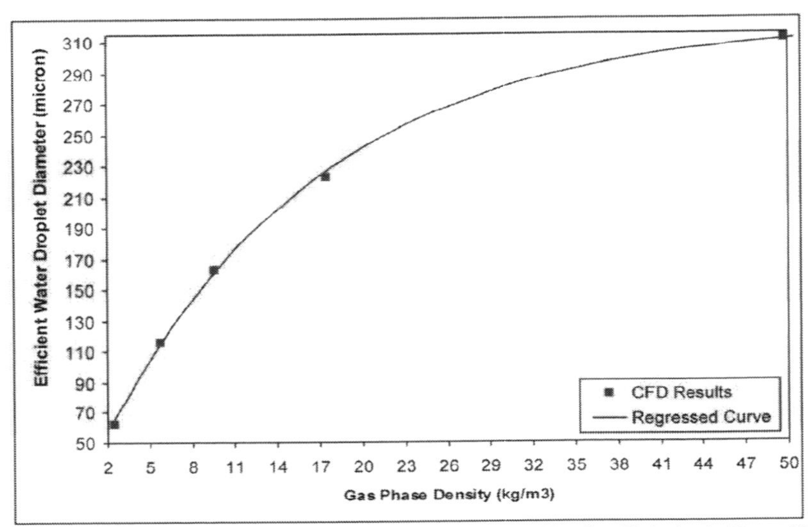

Figure 5-3. Efficient Water Droplet Diameter for Estimation of Settling Velocity in Vapor Phase versus Vapor Density.

5.2.2 *Liquid-Liquid Separation*

A force balance on the oil droplets rising in the water phase or water droplets settling in the oil phase leads to Equation 5-6 (Green and Perry, 2008):

$$V_{sep} = \sqrt{\frac{4\times10^{-6}\, g d_{eff}\left(\rho_W - \rho_{Oil}\right)}{3C_D\rho_c}} \qquad \textbf{(5-6)}$$

where ρ_W, ρ_{Oil}, and ρ_c are densities of water, oil, and continuous liquid phase (oil or water), respectively, in kg/m^3.

In the classic design methods, the drag coefficient for the flow of rising oil droplets in the water phase or settling water droplets in the oil phase is described using Stokes' law, Equation 5-7 (Green and Perry, 2008):

$$C_D = \frac{24}{\mathrm{Re}_p} = \frac{24\mu_c}{1\times10^{-6}\rho_c d_{eff} V_{sep}} \qquad \textbf{(5-7)}$$

where Re_p is particle Reynolds number, and μ_c is viscosity of continuous liquid phase in *Pa.s*.

Substitution of drag coefficient from Equation 5-7 into Equation 5-6 leads to a straightforward relation for calculation of separation velocity in oil-water systems:

$$V_{sep} = \frac{1\times10^{-12} g d_{eff}^{\ 2}(\rho_W - \rho_{Oil})}{18\mu_c} \qquad \textbf{(5-8)}$$

Note, the use of Equation 5-8 is limited to the laminar settling or rising because Stokes' law cannot be used if Re_p exceeds the upper limit of 0.10 (Green and Perry, 2008). King (2002), who has provided a pertinent and useful survey of the fluid-particle interactions, suggested an even lower Reynolds number of 0.01 as the upper limit for Stokes' law.

An analysis of the CFD simulation results showed that the calculated Re_p exceeded the upper limit in several of the case studies (see Table 5-4). Alternatively, as recommended by King (2002), for a $\mathrm{Re}_p < 2\times10^3$, which would cover all the oilfield case studies, the drag coefficient can be estimated accurately by the Abraham equation, Equation 5-9 (Abraham, 1970):

$$C_D = 0.28\left(1 + \frac{9.06}{\sqrt{\mathrm{Re}_p}}\right)^2 = 0.28\left(1 + 9.06\times10^3\sqrt{\frac{\mu_c}{\rho_c d_{eff} V_{sep}}}\right)^2 \qquad \textbf{(5-9)}$$

Table 5-4 presents the CFD simulation results for the liquid-liquid separation for the oilfield cases. Similar to vapor-liquid separation, a residence time is required for the oil and water droplets to penetrate through the liquid-liquid interface. Table 5-4 shows that as viscosity of oil phase varies from 0.73×10^{-3} *Pa.s* to 0.10 *Pa.s*, the interface residence times vary almost linearly from 3.8 *s* to 97.2 *s* for the oil droplets and from 0.9 *s* to 8.4 *s* for the water droplets.

In Table 5-4, the efficient droplet sizes have been calculated from both Stokes' law and the Abraham equation. A comparison between the results indicates that there is a significant difference that occurs while estimating the efficient oil droplet sizes in all five cases and while estimating the efficient water droplet sizes in three cases. Hence, it can be concluded that the use of Stokes' law for estimation of efficient droplet sizes generally results in totally different values from those estimated when using the Abraham equation.

Figure 5-4 shows the variations in the calculated efficient oil droplet size versus viscosity of the water phase. The data of Figure 5-4 would indicate that a constant-value line, $d_{eff,Oil-Liquid}$ = 597 μm, can represent the CFD simulation results when the Abraham equation is used. Therefore, an efficient oil droplet size of 600 μm can be assumed for the Abraham equation to estimate the oil rising velocity in separator design procedures. Figure 5-4 also shows that the incorrect use of Stokes' law for interpretation of CFD results does lead to a totally different and weak correlation between efficient oil droplet size and the viscosity of water phase.

Table 5-4. The liquid-liquid separation characteristics for different oilfields.

			Oilfield-1	Oilfield-2	Oilfield-3	Oilfield-4	Oilfield-5
Oil Drops	V_{sep} (m/s)		0.03175	0.03002	0.01800	0.01494	0.01344
	$t_{interface}$ (s)		3.8	27.7	24.3	74.5	97.2
	Stokes' Law	d_{eff} (μm)	322.7	345.4	340.4	455.4	440.8
		Re_p	26.99	24.84	28.52	7.76	11.69
	Abraham Equation	d_{eff} (μm)	566.9	591.6	608.4	595.2	622.9
		Re_p	47.41	42.55	50.98	10.14	16.52
Water Drops	V_{sep} (m/s)		0.05266	0.01813	0.006056	0.003249	0.001341
	$t_{interface}$ (s)		0.9	2.7	3.1	4.0	8.4
	Stokes' Law	d_{eff} (μm)	583.9	938.1	1159.8	1450.4	1930.6
		Re_p	32.33	2.69	0.89	0.10	0.02
	Abraham Equation	d_{eff} (μm)	1086.8	1072.8	1235.5	1450.4	1930.6
		Re_p	60.18	3.08	0.95	0.10	0.02

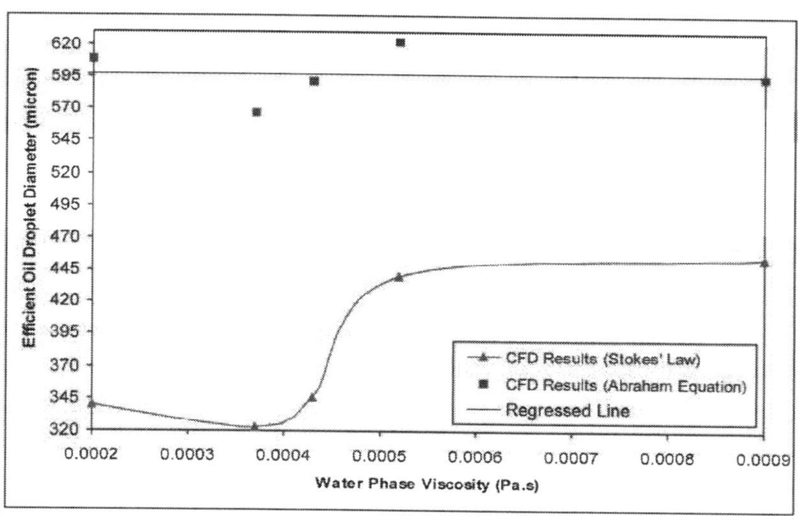

Figure 5-4. Efficient Oil Droplet Diameter for Estimation of Rising Velocity in Water Phase versus Water Viscosity.

It would seem that such a constant value assumed for the Abraham equation may similarly be used in Stokes' law. However, while the maximum variation from the assumed constant value (597 μm) is 5.0% for the Abraham equation; for Stokes' law, the maximum variation from the average value of 381 μm will be 19.5% which would result in errors up to 43% for the estimation of the separation velocities.

Figure 5-5 presents the variations of efficient water droplet size calculated using the Abraham equation versus oil phase viscosity. A linear equation, Equation 5-10, can represent the correlation between efficient water droplet size and oil phase viscosity with corresponding R-squared value of 0.9774.

$$d_{eff,Water-Liquid} = 1095 + 8382\mu_{\text{Oil}} \qquad \textbf{(5-10)}$$

The variation of efficient water droplet size calculated by Stokes' law versus oil viscosity is shown in Figure 5-6. Compared with the Abraham equation based modeling, a much broader range of efficient droplet size as well as a non-linear regression of provided data, Equation 5-11, are required if Stokes' law is used for interpretation of CFD results.

$$d_{eff,Water-Liquid} = 3170\mu_{\text{Oil}}^{0.2253} \qquad \textbf{(5-11)}$$

In summary, Equations 5-6 and 5-9 with the oil droplet size of 600 μm and the water droplet size as per Equation 5-10 can be used iteratively for estimation of liquid-liquid separation velocities. Furthermore, an additional liquid residence time proportional to the oil viscosity of some 100 s for water phase and 10 s for oil phase should be used for separator designs.

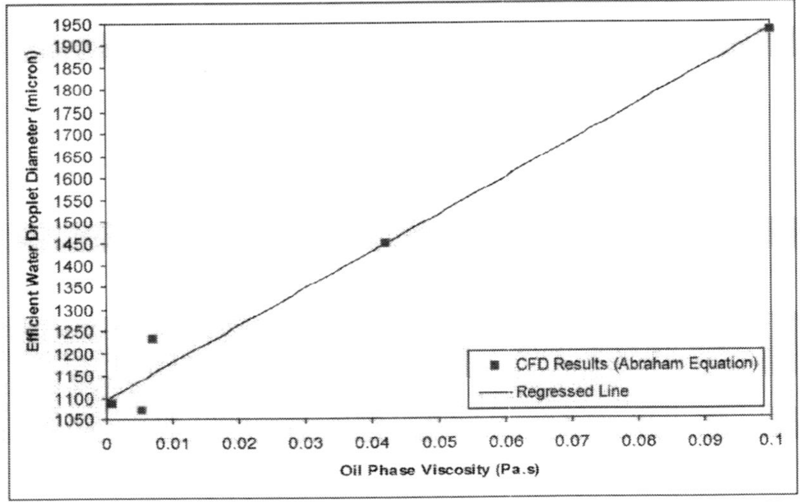

Figure 5-5. Efficient Water Droplet Diameter as a Function of Oil Phase Viscosity to be Used in Abraham Equation.

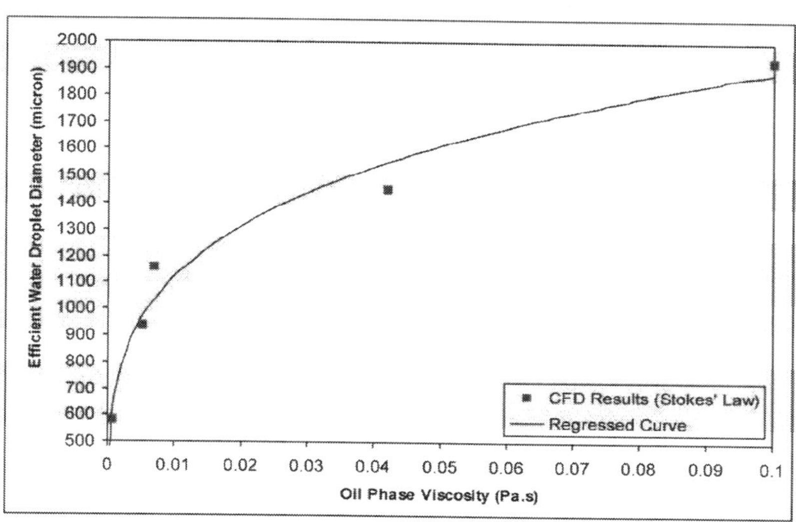

Figure 5-6. Efficient Water Droplet Diameter as a Function of Oil Phase Viscosity to be Used in Stokes' Law.

5.3 Vessel Orientation Considerations

The influence of continuous phase on the movement of discrete phase in horizontal and vertical configurations will be explained from a CFD simulation perspective. The emphasis will be placed on estimating the allowable velocities for the continuous phases.

5.3.1 *Horizontal Arrangement*

Using the two-phase separation models (Figure 5-1) and performing CFD simulations of phase separation at various continuous phase velocities did confirm that oil and water droplets follow the continuous phase flow in their horizontal movements. In other words, the mass averaged horizontal velocities of oil and water droplets as simulated were almost identical to the continuous phase velocities. Therefore, in the iterative design procedure proposed by Svrcek and Monnery (1993), the residence length (L) can be set using the separation height (H_{sep}), the separation velocity (V_{sep}), the continuous phase velocity (V_c) and the required interface residence time ($t_{interface}$) as per Equation 5-12:

$$L \geq \left(\frac{H_{sep}}{V_{sep}} + t_{interface} \right) V_c \qquad \textbf{(5-12)}$$

Note, although the continuous phase can flow with a velocity that carries droplets up to outlet vicinity, these high velocities will lead to re-entrainment of droplets by the continuous phase. The other limit is due to economic considerations in which separators with aspect ratios higher than 9 are rarely economical (Smith, 1987). Thus, there is an upper limit for the continuous phase velocity and the separator aspect ratio.

5.3.2 *Vertical Arrangement*

In order to investigate how changing the separator orientation from horizontal to vertical may influence the phase separation process, CFD simulations were performed using the models represented by Figure 5-7. Similar to the horizontal arrangement, CFD simulations confirmed that the movement of oil and water droplets is completely influenced by the flow of continuous phase. In the vertical configuration, the continuous phase is flowing in the opposite direction to the settling or rising droplets, hence the apparent separation velocities are lower than those in the horizontal case, where forces do not directly oppose one another. However, the relative separation velocities were almost the same as the separation velocities predicted for horizontal arrangement by the CFD simulations. From a practical point of view, the drag force exerted on droplets is a function of the relative velocities (not the apparent velocities). For instance, the drag force on a particle fixed in space with a lighter fluid moving upward is almost the same as the drag force on the particle freely settling in the stationary fluid at the same relative velocity (Green and Perry, 2008). Therefore, the separation velocities and relationships for the horizontal arrangement can also be used for designing vertical separators. However, the apparent separation velocity for a vertical separator will be a function of continuous phase velocity, Equation 5-13:

$$V_{sep,vertical} = V_{sep} - V_c \qquad \textbf{(5-13)}$$

In designing vertical separators, Equation 5-13 should be combined with the vapor-liquid and liquid-liquid separation relationships provided for the horizontal separator models. The most important differences are presented in the following two subsections.

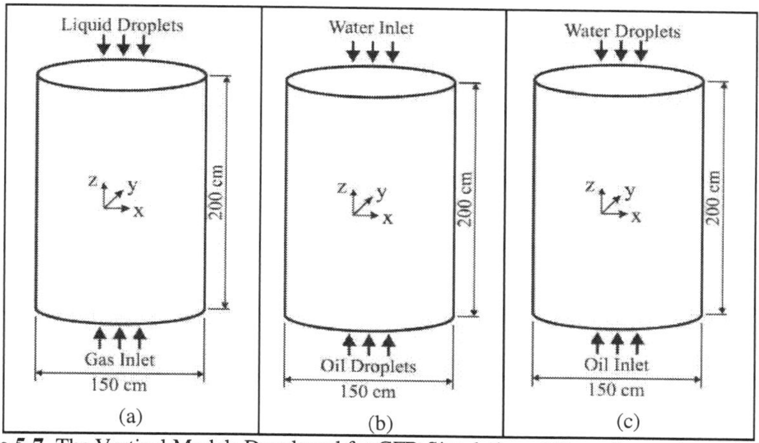

Figure 5-7. The Vertical Models Developed for CFD Simulation of Phase Separation; (a) Vapor-Liquid Separation, (b) Separation of Water Droplets from Oil Phase, and (c) Separation of Oil Droplets from Water Phase.

5.3.2.1 Vapor-Liquid Separation

Separation velocity of oil droplets, V_{sep}, is calculated using Equations 5-1 to 5-4. The calculated value is an upper limit for the vapor velocity at which oil droplets would be suspended in the vapor phase. Hence, according to Monnery and Svrcek (1994), a vapor velocity of around 75% of V_{sep} may be assumed for a realistic design. Note, with this assumption, the apparent settling velocity of oil droplets would be 25% of V_{sep}. Having set the vapor velocity, the internal diameter of separator can be initially determined.

5.3.2.2 Liquid-Liquid Separation

To start the sizing calculation, the separation velocities of oil and water droplets are calculated using Equations 5-6, 5-9 and with the oil droplet size of 600 μm and the water droplet size as per Equation 5-10. These are the maximum apparent separation velocities that may occur in a vertical separator with a diameter of infinity. Since the initial diameter of separator has been set based on vapor-liquid separation requirements, superficial velocities of oil and water phases can be calculated. These values are used in Equation 5-13 to calculate the apparent separation velocities for liquid phases. Furthermore, similar to the horizontal separator, an extra appropriate liquid residence time should also be added. The assumed separator diameter may need to increase if the provided liquid residence time is not sufficient for the required separation. Thus, some iterative calculations are required to specify the vessel diameter.

Alternatively, the following equivalent approach is proposed for a vertical separator configuration. Assume that liquid phase (oil or water) height (H) and volumetric flow-rate (Q), and interface residence time ($t_{interface}$) have been given or set, and the surface area (A) required for liquid-liquid separation is to be calculated. Using the widely accepted design procedures of Monnery and Svrcek (1994), Equation 5-14 should be satisfied:

$$\frac{HA}{Q} = \frac{H}{V_{sep}} + t_{interface} \quad \textbf{(5-14)}$$

Solving Equation 5-14 for "A" results in Equation 5-15:

$$A = \left(\frac{1}{V_{sep}} + \frac{t_{interface}}{H}\right)Q \quad \textbf{(5-15)}$$

Equation 5-15 provides the minimum surface area required for efficient liquid-liquid separation and should be satisfied for both the oil phase and water phase. For the water phase, calculation of required area leads to a straightforward calculation of the required diameter. For the oil phase, however, if there is a baffle plate, the cross-sectional area of down-comer should also be added to the required area. After calculation of the two required diameters for water and oil phases, the values will be compared with the diameter assumed for efficient vapor-liquid separation. The largest value will be selected as the required internal diameter for the vertical separator.

5.3.3 *The Proposed Approach in a Vector Space*

This subsection summarizes the approach proposed in section 5.3 using a vector space. CFD simulation results show that the apparent velocity vector of droplets is the sum of separation velocity vector and continuous phase velocity vector independent of the separator orientation:

$$\vec{V}_{apparent} = \vec{V}_{sep} + \vec{V}_{c} \qquad \textbf{(5-16)}$$

Note, Ishii and Zuber (1979) have also employed the same vector space approach when using the relative settling or rising velocities to present their relationships for various dispersed phase-continuous phase fluid regimes.

Therefore, as concisely demonstrated in subsection 4.2.6.3, the simple models can be used for estimation of realistic separation velocities for both horizontal and vertical arrangements, but efficient droplet sizes, proposed in section 5.2, should be estimated and used with these models.

5.4 Designing the Realistic Optimum Separator for Gullfaks-A

The aim of this section of Chapter Five is to provide the optimum separator design for processing Gullfaks-A (Hansen et al., 1993). For realistic design purposes, the results of the CFD simulations presented in previous sections have been used, and both horizontal and vertical arrangements have been considered. The performance of designed separators has been verified based on the CFD simulations. The supplementary package of the book provides CFD results in terms of pressure, velocity, and density profiles for the resultant fluid flows.

5.4.1 *Design Phase*

The algorithmic design method proposed by Monnery and Svrcek (1994) was used. Their design procedure, however, used the "separation velocities" and interface residence times obtained from CFD simulations to design the most economical separator. The physical parameters for the fluids in the Gullfaks-A separator, presented in Table 3-8, have been taken from Hansen et al. (1993). The maximum flow-rates of the produced gas, oil, and water phases were used in the design. Consequently, a "simple" separator with a diameter of 2.75 *m* and length of 19.40 *m* and an approximate weight of 135270 *kg* was calculated as the most economical design. A vertical separator with a diameter of 10.8 *m* and a height of 16.2 *m* and an approximate weight of 3014660 *kg* turned out to be the most expensive design.

The estimated weights corresponding to all design types are presented in Table 5-5. Based on the volumetric flow-rates and calculated separator dimensions, the "boot" design was excluded from the feasible design set. Due to the large volumetric flow-rate of water, the boot diameter (7.67 *m*) was calculated to be very large compared with the vessel diameter (3.06 *m*). Note, the "boot" design is typically used when the volumetric flow-rate of heavy liquid is not high (Monnery and Svrcek, 1994). Figure 5-8 presents the dimensions of the feasible horizontal separators designed for processing Gullfaks-A.

Because of its large diameter (10.8 *m*) and very large weight (12.8 times heavier than the heaviest horizontal separator, Table 5-5), the designed vertical arrangement is not economical and was excluded from further CFD performance validations. However, in order to study the separation efficiency of the vertical arrangement, the production capacity of Gullfaks-A was assumed to be 25% of the original values, and an additional smaller vertical separator was designed. Figure 5-9 provides the dimensions of the smaller vertical separator designed for processing Gullfaks-A at 25% of total capacity.

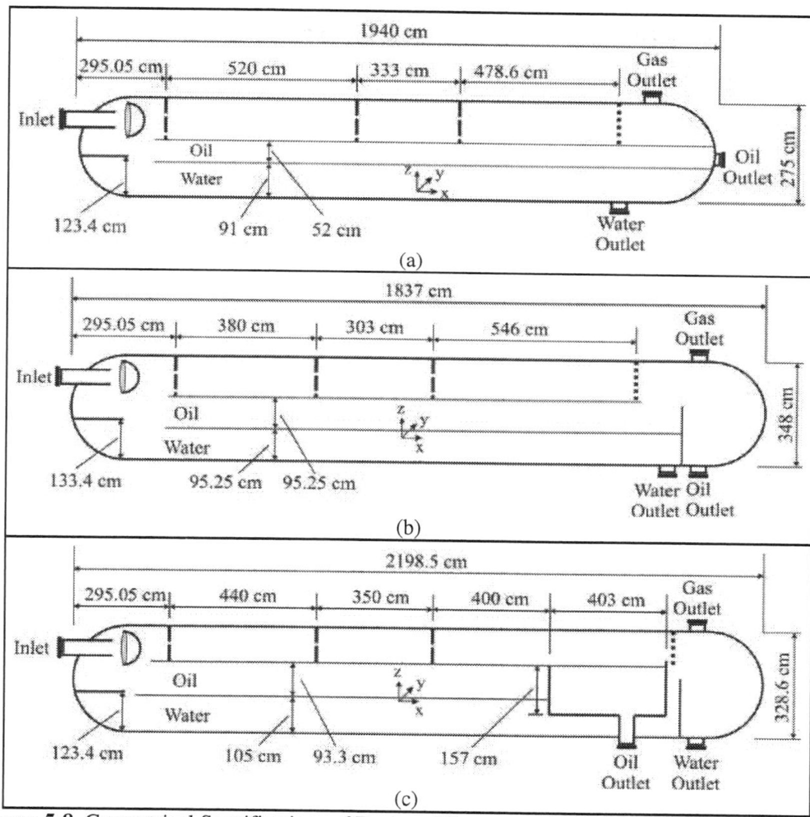

Figure 5-8. Geometrical Specifications of Designed Separators for Gullfaks-A; (a) Simple Type, (b) Weir Type, and (c) Bucket and Weir Type.

Table 5-5. The approximate separator weights calculated for various types of separators processing Gullfaks-A.

Various Designs	Simple	Boot	Weir	Bucket & Weir	Vertical
Approximate Weight (*kg*)	135270	181740	228720	234500	3014660

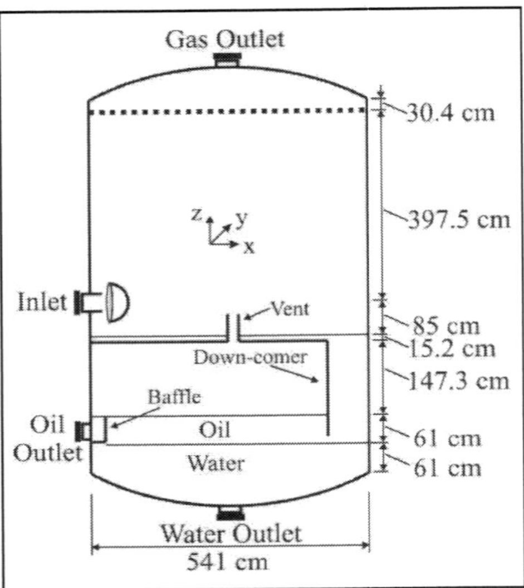

Figure 5-9. Geometrical Specifications of the Vertical Separator Designed for Gullfaks-A (25% Capacity).

5.4.2 *Performance Validations via CFD Simulations*

The separation performance of all the feasible horizontal separators as well as the vertical separator was validated using CFD simulations. Using the developed procedure outlined in Chapter Three for simulation of the original separator, the generation of the grid system was completed for all separators in the Gambit and Fluent environments (Gambit 2.4.6, 2006; Fluent 6.3.26, 2006). Figure 5-10 to Figure 5-12 provide additional detail for the generated mesh systems. The global quality of the produced mesh systems in terms of number of cells, maximum cell squish, cell skewness, and maximum aspect ratio are presented in Table 5-6. Table 5-6 also indicates that a negligible fraction of cells, generally less than 0.08%, are of poor quality. Having set all the CFD parameters for the 1988 or future production conditions, the number of iterations required for convergence varied from 6000 (for "simple" and "weir" designs) to 12000 (for "bucket and weir" design and vertical arrangement). Each iteration takes about 27 *s* on a Pentium D (3.20 GHz) and 2.00 GB of RAM PC. Therefore, a PC run-time of around 45 *h* (for "simple" and "weir" designs) or 90 h (for "bucket and weir" design and vertical arrangement) was required for solution of continuous-phase flow regimes. Note, the iterative process was stopped regularly to check the interface levels and correct them (if necessary). Also, an additional PC run-time of around 3 h per each case study was required for simulation of interactions between the liquid droplets and continuous phases.

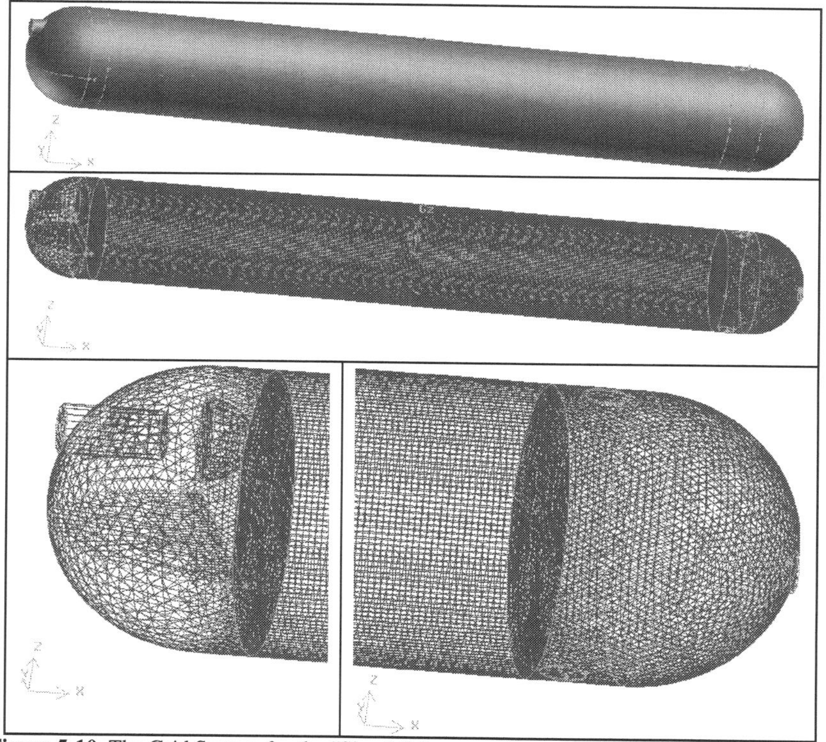

Figure 5-10. The Grid System for the "Simple" Separator Designed for Processing Gullfaks-A Generated in Gambit Environment.

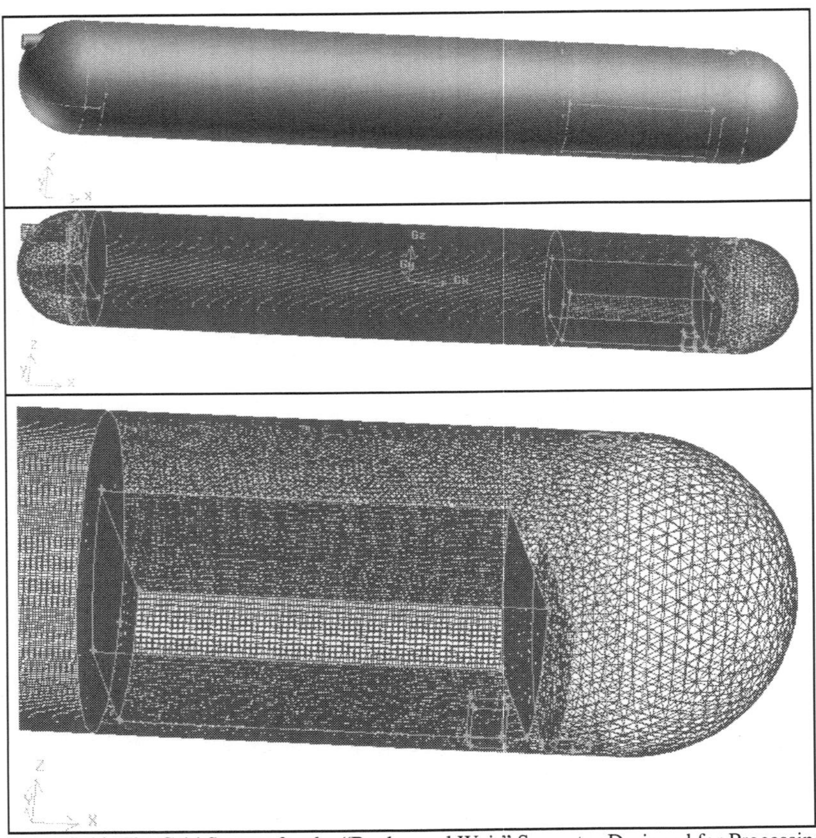

Figure 5-11. The Grid System for the "Bucket and Weir" Separator Designed for Processing Gullfaks-A Generated in Gambit Environment.

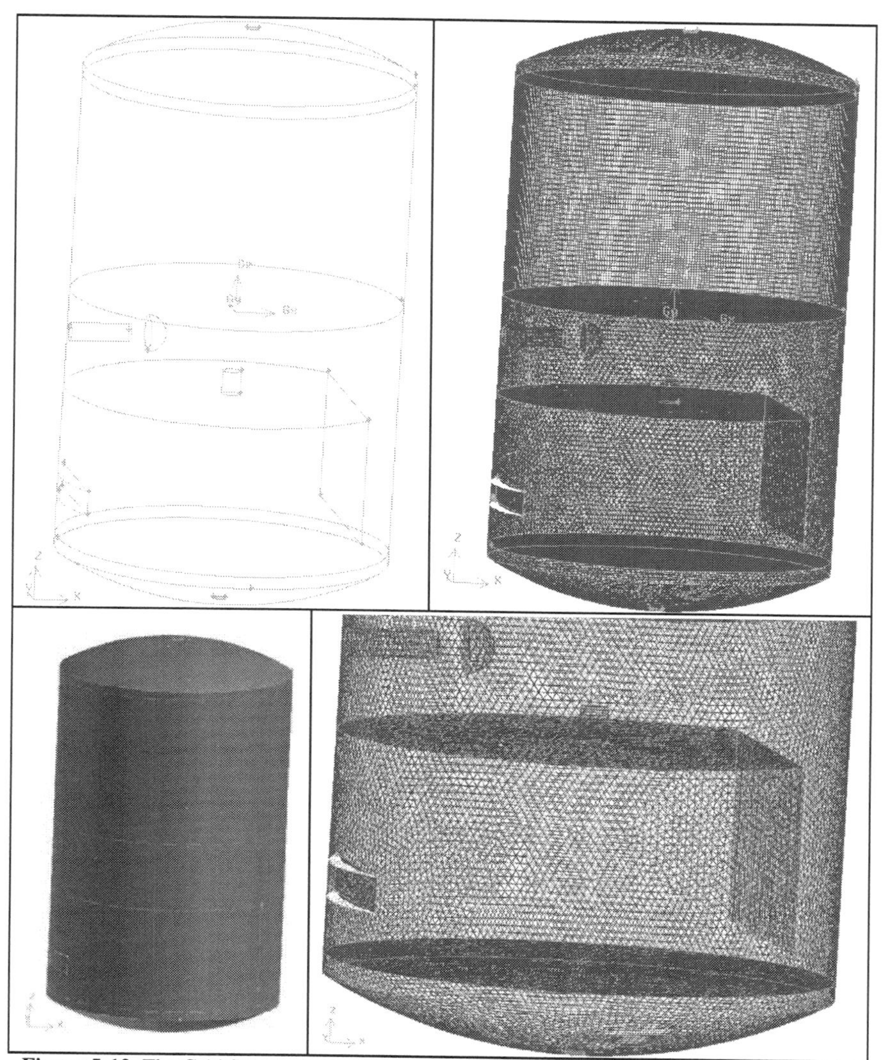

Figure 5-12. The Grid System for the Vertical Separator Designed for Processing Gullfaks-A (25% Capacity) Generated in Gambit Environment.

Table 5-6. Quality of the mesh produced in Gambit environment for various separators processing Gullfaks-A.

<table>
<tr><th colspan="2"></th><th>Number of Cells</th><th colspan="2">Maximum Squish</th><th colspan="2">Maximum Skewness</th><th>Maximum Aspect Ratio</th></tr>
<tr><td colspan="2">Weir</td><td>985834</td><td colspan="2">0.735457</td><td colspan="2">0.895869</td><td>44.4761</td></tr>
<tr><td colspan="2">Simple</td><td>920205</td><td colspan="2">0. 745314</td><td colspan="2">0. 907619</td><td>42.7887</td></tr>
<tr><td colspan="2">Bucket and Weir</td><td>1019479</td><td colspan="2">0. 829993</td><td colspan="2">0. 895890</td><td>44.0854</td></tr>
<tr><td colspan="2">Vertical</td><td>938765</td><td colspan="2">0. 997151</td><td colspan="2">0. 999998</td><td>1108.15</td></tr>
<tr><td colspan="8">Skewness of the produced mesh</td></tr>
<tr><td colspan="2">Skewness Range</td><td>0-0.20</td><td>0.20-0.40</td><td colspan="2">0.40-0.60</td><td>0.60-0.80</td><td>0.80-1.0</td></tr>
<tr><td rowspan="4">Density of Cells</td><td>Weir</td><td>80.1525%</td><td>14.9260%</td><td colspan="2">3.4968%</td><td>1.4233%</td><td>0.0014%</td></tr>
<tr><td>Simple</td><td>72.6225%</td><td>20.5087%</td><td colspan="2">5.6634%</td><td>1.2040%</td><td>0.0014%</td></tr>
<tr><td>Bucket & Weir</td><td>72.5747%</td><td>19.0717%</td><td colspan="2">5.8814%</td><td>2.4626%</td><td>0.0096%</td></tr>
<tr><td>Vertical</td><td>36.7367%</td><td>43.2999%</td><td colspan="2">17.0732%</td><td>2.8129%</td><td>0.0773%</td></tr>
</table>

The macroscopic features of CFD simulations consisting of the pressure, velocity and density profiles for the fluid flows on the central xz-plane of the separators are included in the supplementary package of the book. As was expected, in all separators, the pressure drops assigned to the baffles and demister are low and hence as expected. The velocity vectors and density contours are also as expected.

Table 5-7 presents the separation efficiencies for all the designed separators. Based on the data of Table 5-7, the following comments can be made:

1- From a separation efficiency point of view, "bucket and weir" design is superior to other horizontal designs.

2- Although water separation efficiency is high enough for almost all designs, some horizontal designs suffer from low oil separation efficiencies with the projected increase in the water flow-rate.

3- Separation efficiencies are very high (about 100%) for both the oil and water droplets in the vertical separator.

As noted, the separation efficiencies for horizontal case studies (particularly, for the weir design) are not high enough to ensure complete phase separation. To further study this issue, additional CFD simulations were performed using the available model for "weir" design for the future production condition.

Table 5-7. Separation efficiencies for various separators processing Gullfaks-A.

		Oil Droplets	Water Droplets
Weir	1988	100%	100%
	Future	68.0%	100%
Simple	1988	99.8%	81.3%
	Future	84.0%	100%
Bucket and Weir	1988	98.4%	96.6%
	Future	99.9%	96.5%
Vertical	1988	100%	100%
	Future	100%	100%

In all of the case studies, several parallel xy-planes were defined inside the vessel to record the characteristics of the droplets passing through them. Software was developed to do an analysis on the database provided by the recording surfaces. This analysis did confirm the values shown in Table 5-4 for the separation velocities. However, the captured data also showed that some oil droplets after reaching the oil-water interface were not able to pass through. Further studies indicated that these droplets, while bouncing near the interface along the separator, were carried out by the water phase. Therefore, the existence of "re-entrainment" phenomenon, in which a fraction of oil droplets are re-entrained by the water phase at high water velocities, was observed as part of the CFD simulations.

To further study this phenomenon, additional case studies were performed at various oil and water flow-rates using the same CFD model. Based on the production conditions at 1988 and the future, the oil and water flow-rates were varied from their minimum values to their maximum values in 10% increments. For each case study, the oil and water droplets were injected and tracked to determine if they were carried out by the other liquid phase. The results of CFD simulations indicated that independent of the oil flow-rate value, re-entrainment by water phase occurred at a water flow-rate of greater than 0.1595 m^3/s. Hence, the previous lower water flow-rate of 0.1329 m^3/s was assumed as the maximum value that would avoid re-entrainment of the oil droplets. Using the separator dimensions, the maximum water velocity was calculated to be 0.063 m/s. Based on the maximum water production rate, a minimum cross-sectional area of 5.485 m^2 was calculated for water phase in order to avoid re-entrainment of the oil droplets.

Using the data provided by the CFD simulations, the "stabilized" optimum separator was then designed using the procedure of Monnery and Svrcek (1994), and the performance of this separator was checked using CFD simulations. Figure 5-13 presents the dimensions of the "stabilized" separator. The approximate vessel weight was calculated to be 231530 *kg* which is some 1.23% higher than the non-stabilized separator design shown in Figure 5-8b.

Table 5-8 presents the quality of the grid system produced in the Gambit environment and does indicate that an absolutely negligible fraction of cells, around 0.0015%, is of poor quality. Figure 5-14 provides additional features of the generated mesh system, and the supplementary package of the book includes the detailed fluid flow profiles. As expected, a very high separation efficiency of 100% was calculated for the oil and water droplets for the stabilized separator.

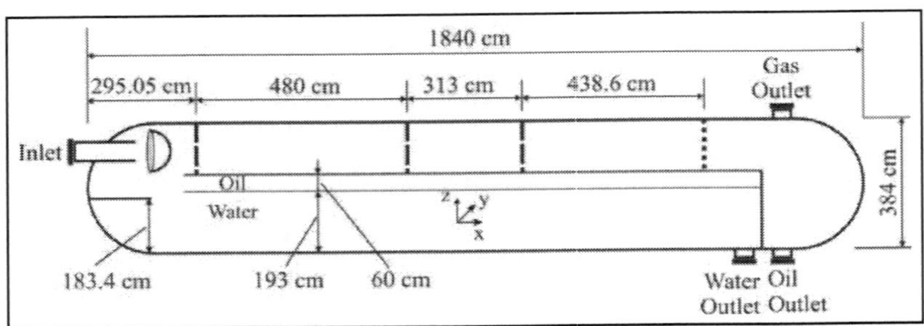

Figure 5-13. Geometrical Specifications of "Stabilized" Separator Designed for Gullfaks-A.

Table 5-8. Quality of the mesh produced in Gambit environment for the "stabilized" Gullfaks-A separator.

Number of Cells	Maximum Squish	Maximum Skewness	Maximum Aspect Ratio
966574	0.746627	0.895882	44.5911

Skewness of the Produced Mesh					
Skewness Range	0-0.20	0.20-0.40	0.40-0.60	0.60-0.80	0.80-1.0
Density of Cells	79.8406%	15.0993%	3.5702%	1.4884%	0.0015%

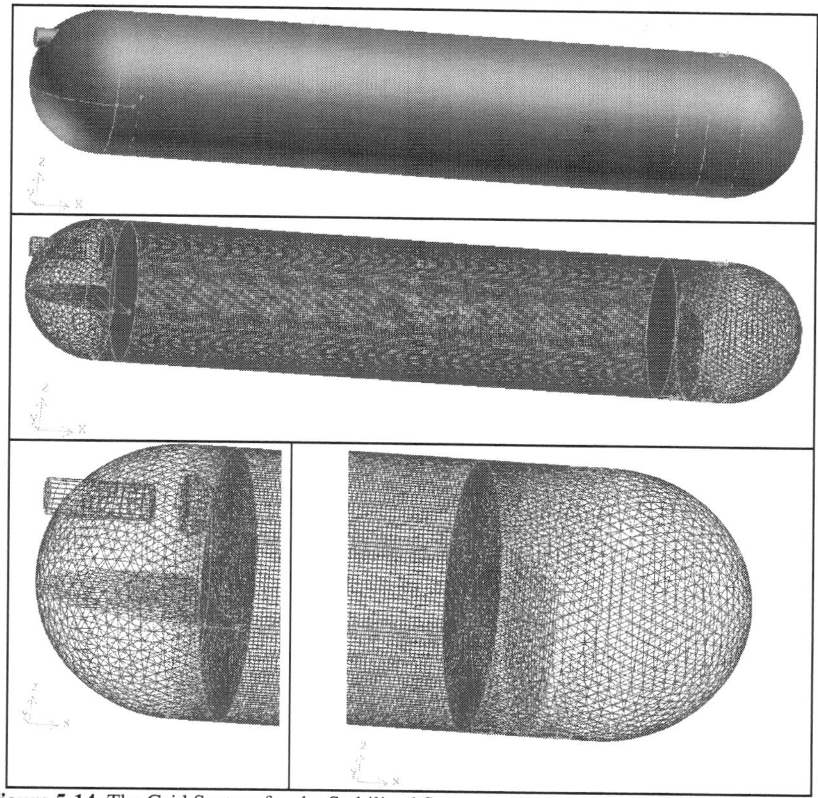

Figure 5-14. The Grid System for the Stabilized Separator Designed for Gullfaks-A Generated in Gambit Environment.

In summary, three different separators have been redesigned and proposed for Gullfaks-A: The first separator design, labeled "redesigned" separator in Chapter Four, was based on the CFD simulation results and the oilfield experiences. The originally installed separator had a reasonable separation efficiency of 98.0% for the 1988 production conditions. Hence, it was assumed that the provided residence times of continuous phases in 1988 might be necessary for a new workable design. Using the maximum fluid flow-rates, the separator was redesigned. The separation performance of this separator, as demonstrated by CFD simulations, was acceptable, but the design suffered from being oversized. An interesting fact emerged from the CFD simulation of the "redesigned" separator was connected with the water phase re-entrainment phenomenon. The "redesigned" separator, interestingly, has the same cross-sectional area of 5.485 m^2 for water phase (suggested at above in the present subsection).

The second separator, labeled "weir" design, was designed based on the data provided by CFD simulations and the two-phase models. Although this separator was smaller than the "redesigned" separator, its separation efficiency was also lower (see Table 5-9). Further CFD studies indicated that this minor inefficiency was caused by water phase re-entrainment of oil.

Using both the data of two-phase models and the findings of the present studies, the third separator, labeled "stabilized" separator, was designed for Gullfaks-A. This design took advantage of all phase separation data provided by CFD simulations as well as the logical optimization methodology presented by Monnery and Svrcek (1994). The separation performance of the "stabilized" separator and its smaller dimensions compared with those of the "redesigned" separator confirmed that "redesigned" separator was oversized and a more economical separator that can accomplish the separation task of Gullfaks-A could be designed. For comparison purposes, Table 5-9 shows the key dimensions and separation efficiencies for the original, "redesigned", "weir" and "stabilized" separators proposed for Gullfaks-A.

Table 5-9. The key geometric specifications and separation efficiencies for various designs proposed for Gullfaks-A.

Design	Geometric Specifications					Separation Efficiency	
	D (m)	L (m)	Layer Thickness (m)		Gravity Separation Length (m)	1988	Future
			Oil Phase	Water Phase			
Original	3.328	16.301	1.0390	0.6250	10.216	98.0%	70.4%
Redesigned	4.845	32.081	1.6300	1.6300	25.830	99.1%	99.7%
Weir	3.48	18.37	0.9525	0.9525	12.290	100%	90.4%
Stabilized	3.84	18.40	0.6000	1.9300	12.316	100%	100%

5.5 Re-Entrainment Constraints

As noted in subsection 5.3.1 and demonstrated in subsection 5.4.2, continuous-phase high velocities can lead to re-entrainment of droplets in horizontal separators. Note, based on the developed design procedure for the vertical arrangement, continuous phase velocities are always lower than droplet separation velocities, and re-entrainment is not an issue with vertical separators. Thus, focusing on horizontal separators, a comprehensive CFD-based study was carried out on the velocity constraints caused by re-entrainment. For this purpose, the developed large-scale CFD models for different horizontal designs (subsection 5.4.1) were used. The simulations were performed for all the oilfield conditions (Table 5-1), and the results were analyzed to see if some general rules could be proposed for the minimization of re-entrainment.

5.5.1 *Vapor-Liquid Re-entrainment*

Since the vapor-liquid separation compartment is essentially the same for the various horizontal designs, only the "weir" designed separator developed for Gullfaks-A was used as the CFD model. The oil and water flow-rates were set as per Gullfaks-A oilfield at the 1988 production conditions. The vapor flow-rate was gradually changed from the normal flow-rate of 1988 in increments of 0.2278 m^3/s (corresponding to vapor velocity increments of 0.054 m/s) to determine the vapor velocity at which the liquid droplets can be re-entrained by the vapor phase from the vapor-liquid interface vicinity. To investigate the issue, several parallel xy-planes were defined inside the vessel to record the characteristics of the droplets passing through them, and the developed software was used to analyze the database provided by the sampling planes. The maximum vapor phase velocities predicted for various oilfields are reported in Table 5-10. Although the aim of current study was not to study the re-entrainment phenomenon in detail, the data of Table 5-10 does indicate that high vapor densities and high oil viscosities reduce the maximum allowable velocity of vapor phase. This trend is also in line with practical experience that indicates high pressures and high oil viscosities limit the allowable vapor velocity, and hence, reduce the gas capacity of the separator (Viles, 1993).

Viles (1993) proposed a method for estimation of the maximum vapor phase velocity. The method was, in fact, the explicit representation of the experimental correlations of Ishii and

Grolmes (1975) for predicting the onset of droplet re-entrainment. In the experimental study, concurrent two-phase fluid flow systems composed of water-nitrogen or water-helium inside a small-scale rectangular transparent apparatus with dimensions of 0.00317 m × 0.0254 m × 0.762 m were used. To specify which equation should be used, an interfacial viscosity number (N_{μ}) and the Reynolds number for surface liquid (oil phase) are evaluated. The interfacial viscosity number is evaluated by Equation 5-17:

$$N_{\mu} = \frac{\mu_{Oil}}{\sqrt{\rho_{Oil}\sigma_{Oil-Gas}\sqrt{\frac{\sigma_{Oil-Gas}}{g(\rho_{Oil}-\rho_{V})}}}} \qquad \textbf{(5-17)}$$

where, μ_{Oil} is oil viscosity in *Pa.s*, ρ_{Oil} and ρ_{V} are densities of oil and vapor phases, respectively, in *kg/m*^{3}, $\sigma_{Oil-Gas}$ is oil-gas surface tension in *N/m*, and g is gravity acceleration in *m/s*^{2}. For evaluation of the surface liquid Reynolds number, physical properties of oil phase and hydraulic diameter of the total liquid phase (oil and water) are used. For the region of rough turbulent flow regimes (Re_{Oil}>1635), which is the case for all the various oilfield case studies, the maximum vapor phase velocity ($V_{V,\max}$) is estimated by Equation 5-18 (Viles, 1993):

$$V_{V,\max} = \begin{cases} N_{\mu}^{0.8}\dfrac{\sigma_{Oil-Gas}}{\mu_{Oil}}\sqrt{\dfrac{\rho_{Oil}}{\rho_{V}}} & ; \; N_{\mu} \le 0.0667 \\ 0.1146\dfrac{\sigma_{Oil-Gas}}{\mu_{Oil}}\sqrt{\dfrac{\rho_{Oil}}{\rho_{V}}} & ; \; N_{\mu} > 0.0667 \end{cases} \qquad \textbf{(5-18)}$$

Table 5-10. The maximum safe velocities of vapor phase determined for the various oilfield case studies.

		Oilfield-1	Oilfield-2	Oilfield-3	Oilfield-4	Oilfield-5
Vapor Density (*kg/m*^{3})		17.46	49.7	9.50	5.70	2.44
Oil Viscosity (*Pa.s*)		0.73×10^{-3}	5.25×10^{-3}	6.90×10^{-3}	42.0×10^{-3}	100×10^{-3}
$U_{V,\max}$ (*m/s*)	CFD Predictions	1.63	0.54	1.85	1.14	1.09
	Estimated by Equation 5-18	2.62	1.06	2.36	0.82	0.54

Based on Equation 5-18 and the available oilfield data, Viles (1993) has presented some general guidelines for the common case of rough turbulent flow regimes. The guideline for the high-pressure operating region ($P > 6900\ kPa$) is to use a vessel aspect ratio of less than 5 to avoid droplet re-entrainment. Note, higher aspect ratios may be used at lower pressures. Care should also be taken for heavy oil (<30°*API*) separator sizing as re-entrainment becomes more likely as oil viscosity increases. For such cases, increasing the operating temperature of the separator in order to reduce the oil viscosity has been suggested (Viles, 1993).

Table 5-10 presents maximum safe velocities for several oilfield case studies as predicted by Equation 5-18. Comparison of these results with the corresponding CFD predictions (Figure 5-15) would indicate that some of the CFD simulation data are underestimated and some are overestimated by Equation 5-18. Therefore, it is concluded that the methodology proposed by Viles (1993) and the resultant guidelines are only partially confirmed by the CFD simulations.

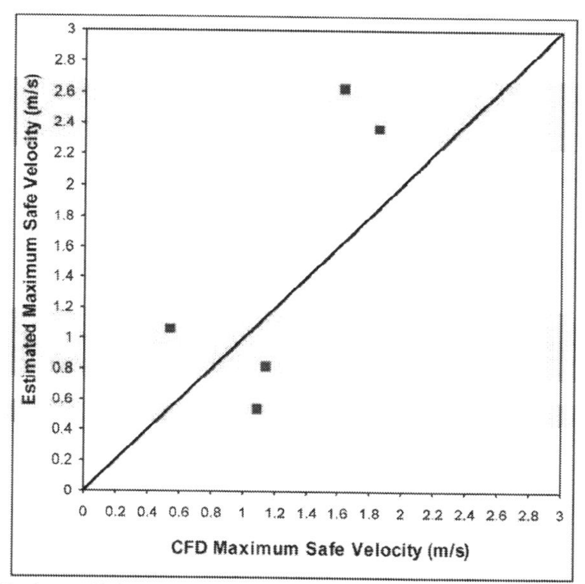

Figure 5-15. Maximum Safe Velocity for the Vapor Phase Estimated by Equation 5-18 versus the Corresponding CFD Predictions.

5.5.2 *Liquid-Liquid Re-entrainment*

The geometry of the liquid-liquid separation compartment is often different for horizontal separator design configurations. Thus, all the horizontal vessel designs except for "boot" design were part of the CFD simulation case studies. The reason why the "boot" design has been excluded from the study will be explained later in this subsection.

The CFD models are those developed for Gullfaks-A oil production platform (Subsection 5.4.1). The vapor flow-rate was set to the Gullfaks-A oilfield rate, and the oil and water flow-rates were gradually varied from their minimum values in increments of 0.02658 m^3/s which corresponded to velocity increments of 0.0083 m/s and 0.0126 m/s for oil and water phases, respectively. For each case study, having set all the necessary CFD parameters and obtaining the converged solution for continuous phases, oil and water droplets were injected and tracked to see if they were re-entrained by the other liquid phase. Again, parallel xy-planes were defined in the liquid-liquid interface vicinity (sampling planes) providing data on the passing droplets. These data were analyzed using the developed software.

The results of CFD simulations using "simple", "weir", and "bucket and weir" models at selected oilfield conditions indicated that the oil phase does not re-entrain the water droplets even at a very high velocity of 1.11 m/s. Therefore, the focus was placed on studying re-entrainment by water phase. Hence, the "boot" design was excluded from the study since the flow-rate of water phase is typically very low in this design and cannot re-entrain the oil droplets.

The CFD studies of the "simple" and "weir" separators showed that re-entrainment by water phase might strongly influence the separation efficiency in these designs. Table 5-11 presents the maximum safe cross-sectional velocities for the "simple" and "weir" designs predicted by the CFD simulations. The data of Table 5-11 do show that high oil viscosities reduce the maximum safe velocity of water phase. Figure 5-16 shows the variations in the maximum safe velocities of the water phase versus viscosity of the oil phase for both "simple" and "weir" designs.

Table 5-11. The maximum safe velocities of water phase in "simple" and "weir" designs predicted for various oilfield case studies.

		Oilfield-1	Oilfield-2	Oilfield-3	Oilfield-4	Oilfield-5
Oil Viscosity (*Pa.s*)		0.73×10^{-3}	5.25×10^{-3}	6.90×10^{-3}	42.0×10^{-3}	100×10^{-3}
$V_{Water,\max}$ (*m/s*)	Simple Design	0.085	0.093	0.077	0.062	0.031
	Weir Design	0.069	0.063	0.063	0.038	0.019

As shown in Figure 5-16, a linear equation can reasonably represent the CFD-based correlations for each design. Equation 5-19 presents the result of linear regression fit for $V_{Water,\max}$ - μ_{Oil} data for the "simple" design:

$$V_{Water,\max\_simple} = 0.08734 - 0.5696\mu_{\text{Oil}} \qquad \textbf{(5-19)}$$

The R-squared value for this linear regression is 0.9522.

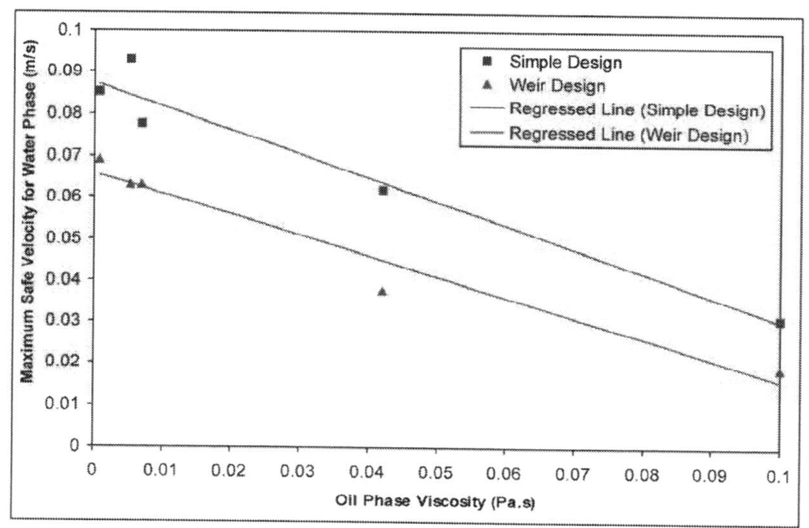

Figure 5-16. Maximum Safe Velocity of Water Phase for "Simple" and "Weir" Designs versus Oil Phase Viscosity.

Similarly, a linear regression fit for $V_{Water,\max}$ - μ_{Oil} data for the "weir" design is presented as Equation 5-20 with corresponding R-squared value of 0.9595:

$$V_{Water,\max\_weir} = 0.065767 - 0.4982\mu_{\text{Oil}} \qquad \textbf{(5-20)}$$

As Figure 5-16 shows, the regressed lines for the "simple" and "weir" designs are approximately parallel to each other. In fact, the $V_{Water,\max}$ values for "weir" design are around 0.018 *m/s* lower than the corresponding $V_{Water,\max}$ values for "simple" design.

The results of CFD simulation for the "bucket and weir" design indicated that, in contrast with the "simple" and "weir" designs, the oil droplets were not re-entrained by the water phase even at a high velocity of 0.82 *m/s*. The obvious reason is the different geometry of water phase compartment in the "bucket and weir" design. To clarify the issue from a CFD point of view, various separation sections of "bucket and weir" design can be compared with those of "simple" and "weir" designs. As shown in Figure 5-8, the gravity separation zones for all three designs are the same. However, after the gravity separation zone, the water phase flows through a complicated and totally different path in the "bucket and weir" design resulting in no virtual path for the oil droplets to move with the water phase. Therefore, oil droplets separate from water phase near the oil weir and become part of the continuous oil phase.

In summary, the CFD simulations did verify that oil droplets may be "re-entrained" by the water phase at a high velocity in horizontal separators. Although significant separation inefficiencies caused by "liquid-liquid" re-entrainment are not likely to be experienced in "boot" and "bucket and weir" designs, a safe upper limit should be empirically determined or estimated by Equations 5-19 and 5-20 for the water phase velocity to avoid "liquid-liquid" re-entrainment in the "simple" and "weir" designs. Therefore, the geometry of water phase compartment in horizontal separators should be considered as a key factor in the liquid-liquid re-entrainment phenomenon.

5.6 Summary

CFD simulations were developed for selected aspects of phase separation. The focus was placed on hydrocarbon-water systems, and the oilfield separator data ranging from light oil conditions to heavy oil conditions were used in these multiphase separator simulations. An efficient combination of two multiphase simulation models available in Fluent, VOF and DPM, with appropriate model assumptions and settings was used. Two independent sets of CFD simulations, one for vapor-liquid separation and the other for liquid-liquid separation, were performed using simple and efficient grid systems.

When compared to classic design strategies, CFD simulations indicated that additional residence times are necessary for droplets to pass through the interfaces. The additional liquid-liquid interface residence times were estimated to be proportional to the oil viscosity, however, 100 *s* for water phase and 10 *s* for oil phase are recommended for separator design.

In the vapor-liquid separation compartment, the efficient droplet size and the appropriate extra vapor residence time (for droplet penetration through the interface) were estimated as a function of the vapor density. It was shown that for the three-phase separator case study, the efficient separation of oil droplets from gas phase results in total separation of water droplets from the gas phase.

For the liquid-liquid separation process, the use of Abraham equation instead of Stokes' law was recommended since the upper limit of Stokes' law was exceeded in several case studies. Using the Abraham equation in liquid-liquid separation calculations, the efficient droplet size was estimated based on continuous phase viscosity. Hence, for water droplets, a linear regression fit based on the oil phase viscosity was developed. An efficient oil droplet size of 597 μm (or simply 600 μm) resulted when the Abraham equation was used for estimation of oil rising velocity in separator design procedures. Furthermore, it was shown that the use of Stokes' law for interpretation of CFD results does lead to a weak correlation between efficient droplet sizes and continuous phase viscosities.

CFD simulations confirmed that the movement of oil and water droplets is governed by the flow of continuous phase. So, in horizontal arrangements, oil and water droplets completely follow the continuous phase flow in the horizontal direction, while in vertical arrangements, apparent separation velocities of oil and water droplets are directly affected by the continuous phase flow.

The relative separation velocities in vertical separators were almost identical to the separation velocities predicted for horizontal separators. Thus, the separation velocities and relationships for the horizontal separators can also be used for designing vertical separators. However, it should be noted that the apparent separation velocity for a vertical separator is equal to the separation velocity in horizontal separator minus the continuous phase velocity in vertical separator. Therefore, apparent separation velocities in horizontal separators are always higher than those in vertical separators.

The algorithmic design method of Monnery and Svrcek (1994) was modified to use CFD simulation results to specify a realistic optimum separator design/size. The performance of the designed separator was also verified using the CFD simulations.

Finally, a comprehensive CFD-based study on the velocity constraints caused by re-entrainment in horizontal separators was carried out. Practical experience and CFD simulations show that high vapor densities and high oil viscosities reduce the maximum safe velocity of the vapor phase. An empirically-based method proposed by Viles (1993) for predicting vapor-liquid re-entrainment was also tested and the results were compared with the CFD results of the current study. This method was based on the empirical study of Ishii and Grolmes (1975) in which only water-nitrogen and water-helium fluid systems in a small-scale rectangular apparatus were used for data generation. Therefore, the equations developed by Viles (1993) could only approximate the predicted CFD results.

The results of CFD simulations, using all the feasible horizontal designs and selected oilfield conditions, indicated that the oil phase does not re-entrain the water droplets, but the oil droplets may be "re-entrained" by the water phase at a high velocity. In the "simple" and "weir" designs, it was observed that re-entrainment by the water phase may strongly influence the separation efficiencies. For these designs, the maximum safe velocity of water phase was regressed as a linear function of the oil viscosity. The regressed lines were almost parallel, and the maximum safe water velocities for the "weir" design were 0.018 *m/s* lower than the corresponding values for the "simple" design.

The "boot" design, because of its very low water phase flow-rate, was excluded from the study. Moreover, CFD simulation performed on "bucket and weir" design indicated that oil droplets are not re-entrained by water phase. Therefore, it was concluded that significant separation inefficiencies caused by "liquid-liquid" re-entrainment are not likely to be experienced in "boot" and "bucket and weir" designs, but a correct upper limit should be assumed for water phase velocity to avoid "liquid-liquid" re-entrainment in the "simple" and "weir" designs. Consequently, the geometry of water phase compartment in horizontal separators is a key factor affecting the liquid-liquid re-entrainment phenomenon.

Chapter Six: Conclusions and Recommendations

This book has provided a detailed approach to the use of CFD as a method for the modeling and simulation of the performance of the multiphase separators. The developed CFD models provided both macroscopic and microscopic understanding of the phase separation phenomenon. Compared with the previous CFD based studies of multiphase separators, the current study does provide realistic strategies for CFD simulation of multiphase separators, and the book does provide all details of developed CFD models. The other significant accomplishments are as follows:

1- Symmetrical fluid flow profiles have generally been assumed in previous studies, and only half of the separator volume has been modeled. However, even the results of these simplified CFD models prove that the plug flow regime cannot be assumed, hence, the assumption of symmetrical fluid flow profiles is not realistic. In this study the total volume of multiphase separators was modeled.

2- The quality of produced computational grid system has not been verified in the previous studies, and coarse grid systems that have been used because of computational restrictions in some studies, would produce poor/doubtful results. The high quality of produced computational grid systems was verified in the present study.

3- In order to reduce the required computational memory and time in some of the previous studies, three-phase fluid flows have been simulated by two-phase CFD models. This approach would reduce the validity of produced CFD profiles and was not used for simulation of three-phase separators in our study. Instead, all fluid phases present in a multiphase separator with their corresponding physical properties were considered in the CFD simulations, thus providing very high quality details of the phase separation features.

4- In the previous studies, only indirect criteria such as the liquid retention time, the volumetric utilization, and the standard deviation of time averaged velocity have in general been used for the evaluation of separation efficiencies. Improving these factors can lead to better plug flow regimes inside separators, but it cannot necessarily lead to increased separation efficiencies. In the current CFD study, to define an effective criterion, separation efficiencies were evaluated directly based on the mass distribution of fine fluid droplets as they were tracked by a suitable multiphase CFD model, DPM. Thus, the realistic performance of the separators was simulated.

5- In the previous studies, CFD based modifications have generally been concentrated on the separator internals such as flow-distributing baffles. However, as demonstrated in this study and previously emphasized by Lyons and Plisga (2005), optimizing the separator internals has only a minor effect on the separator performance, and an essentially inefficient separator cannot become workable simply by optimizing its internals. To overcome this shortcoming, we applied realistic modifications to an operating large-scale separator and also proposed useful and improved design criteria for existing design methods.

We performed three major CFD based studies whose the most significant features and results are presented in the following sections.

6.1 CFD Simulation of Pilot-Plant-Scale Two-Phase Separators

The paper titled "Analytical Study of Liquid/Vapor Separation Efficiency", by Monnery and Svrcek (2000), and a field pilot plant skid at the Prime West East Crossfield gas plant were used as the basis for the CFD model and to provide experimental data. Two approaches were implemented in the CFD simulations. The overall strategy in the first approach, DPM multiphase modeling, was adopted from Newton et al. (2007) and only the vapor-liquid compartment of the separators was simulated. In this approach, the gas-liquid interface was assumed to be a frictionless wall which trapped the droplets coming into contact with it. In the second approach, an effective combination of DPM and VOF multiphase models within Fluent 6.3.26 was used. This CFD model was based on the physics of involved phase separation process and the characteristics of the Fluent multiphase models. To implement the combined VOF-DPM model, having developed a CFD simulation of the overall phase behavior of the fluid flows using the VOF model, droplets of oil and water were injected at the inlet nozzle and tracked by DPM model. The important conclusions that can be drawn from these two approaches are:

- Although the DPM approach was quite successful in predicting the incipient velocities, the produced diagrams for separation efficiency versus gas velocity were not realistic. In fact, modeling only the gas phase compartment was not sufficient to capture all details of the phase separation phenomenon. The approach shortcoming is a result of ignoring the existence of the continuous liquid phase. Thus, gas-liquid interactions and, more significantly, the dynamic interactions between liquid droplets and continuous liquid phase are neglected in the DPM approach.
- Three different mean droplet sizes were tested for prediction of incipient velocities in the DPM approach. The CFD results showed a minor difference in favor of the case of $\overline{d}$ = 150 μm. Thus, the use of this well-known design value for prediction of incipient velocity was validated by the study.
- The VOF-DPM approach was a substantial modification to the DPM approach as the model did include the continuous liquid phase within the calculations. From a practical point of view, the obtained separation efficiency data and diagrams were correct and the poor behavior of the DPM-only approach, such as perfect separation at velocities higher than the incipient velocity, was completely eliminated. There was excellent agreement between simulated phase separation behavior and the experimental data and observations in most of the case studies.
- Based on the obtained fluid flow profiles, all the two-phase separators were shown to essentially operate at a constant pressure.
- In the developed CFD simulations, for the first time, both droplet coalescence and breakup were modeled. The results showed that while droplet coalescence occurred rarely, droplet breakup was a common phenomenon particularly at high velocities. The simulation results do show that higher velocities intensified the number of droplet breakups in horizontal separators while in vertical separators, higher pressures stabilized the number of breakups to a constant rate of about 20%.
- Using the developed C++ software, the size distribution of the droplets exiting through the gas-outlet in different separators was calculated. The results showed that mist eliminators may operate more efficiently in horizontal separators than in vertical separators. In horizontal separators, the particle size distribution was always very narrow and the emerging droplets

were totally dominated by droplets of mean diameter larger than 10 μm, which can be separated by a properly designed demister within reasonable operating conditions and separator capacity. However, the average values of the spread parameter were generally lower for the vertical separator than those for the horizontal separators, and both very fine and very large droplets were still present in the vertical separator gas outlet.

6.2 CFD Based Study of a Large-Scale Three-Phase Separator

A three-phase separator located in the Gullfaks oilfield in the Norwegian sector of the North Sea was simulated. Based on the characteristics of the Fluent multiphase models and the obtained results for the pilot-plant scale separators, the combined VOF-DPM model of Fluent 6.3.26 was implemented. In this study, the installed distribution baffles and mist eliminator were modeled using the Porous Media Model of Fluent which required the available specifications and design information for the three-phase separator. Using the available theoretical approaches and experimental correlations, a useful methodology for estimation of droplet size distribution, which is necessary for implementing DPM approach, was developed. In order to overcome serious separation problems experienced by the projected increase in the water flow-rate and to enhance the separation efficiency, design approaches were also tested using CFD simulations. The significant results and conclusions are presented in the following:

- Compared to the original study of Hansen et al. (1993), the developed model did provide more rational and high-quality details of fluid flow profiles leading to a realistic overall picture of the phase separation in all zones of the separator. Thus, the realistic performance of the separator was simulated and the microscopic features of the three-phase separator were studied in detail.
- The CFD simulations indicated that the droplet breakup with an average rate of 76% was a common phenomenon when the dispersed droplets came into contact with the deflector baffle. Because of these droplet breakups, the volume median diameter of droplets was predicted to decrease to about 67% of its initial value.

- In line with the oilfield experience, the CFD simulations showed that serious separation inefficiencies would be encountered with the existing separator at the projected increase in the flow-rate of produced water.
- To overcome the inefficiencies, minor modifications such as adjusting baffle positions and liquid levels were tested via CFD simulations. However, the results did show that the minor modifications cannot resolve the essential separation inefficiencies in the Gullfaks-A separator. Therefore, the separator was redesigned using the classical design method of Monnery and Svrcek (1994) and implementing appropriate liquid retention times as were implied by the satisfactory performance of the separator at the 1988 production conditions.
- The excellent separation performance, shown in the CFD simulation, of the redesigned separator did confirm that, a realistic optimum separator can be specified if the algorithmic method of Monnery and Svrcek (1994) is modified to use realistic separation velocities.
- It was shown that the popular classic methods, mostly due to a lack of a useable mathematical model for estimation of droplet "separation velocities", do result in a conservative design and would specify extremely oversized separators for Gullfaks-A.
- Sensitivity analyses performed on the developed CFD model for Gullfaks-A separator reconfirmed the high quality of the grid system and robustness of the DPM results with respect to solution repetitions. These analyses also indicated that some uncertainties while defining droplet size distribution or estimating surface tensions had only little effect on the CFD simulation of multiphase separators.
- Although the overall performance of wire mesh demisters, such as pressure and velocity profiles and some separation features, could be properly simulated, the separation efficiency of wire mesh demisters and their operability ranges could not be simulated using the existing CFD models. The CFD models appear to under-predict the demister efficiency. This shortcoming is possibly a result of the complicated separation mechanisms involved with wire mesh demisters.

6.3 Improved Design Criteria

The aim of this phase of the research was to exploit the implemented VOF-DPM model assumptions/settings and provide some generally useful updated phase separator design criteria. These criteria could then be combined with the algorithmic design method proposed by Monnery and Svrcek (1994) to specify a realistic optimum separator. In order to simulate the various aspects of the phase separation phenomenon, oilfield separator data ranging from light oil conditions to heavy oil conditions were used. As a result, a systematic method for estimation of realistic separation velocities was developed. The velocity constraints caused by re-entrainment in horizontal separators were also studied via the comprehensive CFD simulations. The important CFD-based findings are highlighted in the following:

- In contrast with classic design strategies, additional residence times should be assumed for droplet penetration through the interfaces. The appropriate extra vapor and liquid residence times for droplet penetration through the interfaces were shown to be function of the vapor density and the oil viscosity, respectively. For conservative separator designs, some 60 s for vapor phase, 100 s for water phase and 10 s for oil phase can be assumed as the required interface residence times.
- In the liquid-liquid separation compartment, the use of the Abraham equation instead of Stokes' law was recommended because the upper limit of Stokes' law was exceeded in several case studies. Moreover, it was shown that incorrect use of Stokes' law for interpretation of CFD results does lead to a weak correlation between design droplet sizes and continuous phase viscosities.
- The efficient/design droplet size for estimation of settling velocities in vapor and oil phases was shown to be function of the vapor density and oil viscosity, respectively. An oil droplet size of around 600 μm can be assumed if the Abraham equation is used for estimation of the rising velocity of oil droplets out of the water phase.
- CFD simulations do show that the movement of oil and water droplets is completely influenced by the continuous phase flow. Thus, in horizontal arrangements, oil and water droplets are part of the continuous phase flow in their horizontal movements, while for

vertical arrangements, apparent separation velocities of the oil and water droplets are directly affected by the continuous phase flow.

- The relative separation velocities in vertical arrangements are almost identical to the separation velocities predicted for horizontal arrangement. Therefore, the separation velocities and relationships obtained for the horizontal arrangement can be used for designing vertical separators. However, note that the apparent separation velocity for a vertical separator is equal to the separation velocity in horizontal separator minus the continuous phase velocity in vertical separator. Thus, apparent separation velocities in horizontal separators are always higher than those in vertical separators.
- As verified by CFD simulations, the algorithmic method proposed by Monnery and Svrcek (1994) can be modified to use realistic separation velocities to specify an effective optimum separator design/size.
- In line with practical experience, CFD simulations indicated that high vapor densities and high oil viscosities reduce the maximum safe velocity of vapor phase.
- The results of CFD simulations, using all the feasible horizontal designs and various oilfield conditions, indicated that the oil phase does not re-entrain the water droplets. However, it was observed that re-entrainment by water phase may strongly influence the separation efficiencies in the "simple" and "weir" designs. For these designs, the maximum safe cross-sectional velocity of water phase was regressed as a linear function of the oil viscosity.
- The geometry of water phase compartment in horizontal separators was shown to be a key affecting factor in the liquid-liquid re-entrainment phenomenon.

6.4 Recommendations

The performed research project did clearly show the benefits CFD analyses can provide in optimizing the design of new separators and solving problems with existing designs. Consequently, a direct tangible benefit to industry is expected through applying the developed CFD simulation strategies and the established new/improved design criteria. Recommendations for future studies would include the following:

- With expected development in CFD modeling tools, the complicated separation mechanisms involved with wire mesh demisters can be simulated, and the realistic separation performance of wire mesh demisters would be studied.
- As the rate of droplet coalescence has always been predicted to be very low even for the CFD model developed for the wire mesh demister, the validity of the droplet collision model of the Fluent need be investigated using experimental data.
- In the previous CFD studies, indirect criteria such as the fluid flow profiles, the liquid retention time, and the volumetric utilization have been used for improving the position or the configuration of separator internals. As noted, these criteria do not necessarily lead to increased separation efficiencies. Thus, it is recommended that the direct separation efficiency criterion developed in this study be used for the CFD optimization of separator internals.
- Although the developed new/improved design criteria resulted in a realistic optimum separator for the Gullfaks-A case study, it is recommended that these criteria be empirically validated using different oilfield fluids in the large-scale multiphase separators.

References

Abernathy, M.W.N., "Gravity Settlers, Design", in "Unit Operation Handbook", J.J. McKetta (Ed.), Vol. 2, Marcel Dekker, 1993.

Abraham, F.F., "Functional Dependence of Drag Coefficient of a Sphere on Reynolds Number", Physics of Fluids, 13, 1970, 2194-2195.

Anderson, J.D., "Computational Fluid Dynamics, The Basics with Applications", McGraw-Hill, 1995.

Angeli, P., Hewitt, G.F., "Drop Size Distribution in Horizontal Oil-Water Dispersed Flows", Chem. Eng. Sci., 55, 2000, 3133-3143.

Antonoff, G.N., "Surface Tension at the Boundary of Two Layers", J. Chim. Phys., 5, 1907, 372-385.

Arnold, K., Stewart, M., "Surface Production Operations", 3rd Edition, Elsevier, 2008.

Arntzen, R., "Gravity Separator Revamping", Dr.-Ing. Dissertation, Norwegian University of Science and Technology, Trondheim, Norway, 2001.

Austrheim, T., "Experimental Characterization of High-Pressure Natural Gas Scrubbers", PhD Dissertation, University of Bergen, Bergen, Norway, 2006.

Blezard, R.G., Bradburn, J., Clark, J.G., Cohen, D.H., Costaschuk, D., Downie, A.A., Fowler, P., Hassoun, L., Hunt, A.P., Kirton, D., Knight, F.I., Lach, J.R., Law, E.J., McDonald, P.A., Morrison, A.K., Cairney, J.M., Naik, H., Sutton, W.J.E., Thompson, P., "Production Engineering", in "Modern Petroleum Technology", R.A. Dawe (Ed.), 6th Edition, Vol. 1, Institution of Petroleum, John Wiley & Sons, 2000.

Branan, C., "The Process Engineers Pocket Handbook", Vol. 2, Gulf, 1983.

Chin, R.W., Stanbridge, D.I., Schook, R., "Increasing Separation Capacity with New and Proven Technologies", Society of Petroleum Engineers, SPE-77495, 2002, 1-6.

Churchill, S.W., "Viscous Flows, The Practical Use of Theory", Butterworths, 1988.

Coker, A.K., "Ludwig's Applied Process Design for Chemical and Petrochemical Plants", Vol. 1, 4th Edition, Elsevier, 2007.

El-Dessouky, H.T., Alatiqi, I.M., Ettouney, H.M., Al-Deffeeri, N.S., "Performance of Wire Mesh Mist Eliminator", Chemical Engineering and Processing, 39, 2000, 129-139.

Evans, F.L., "Equipment Design Handbook for Refineries and Chemical Plants", Vol. 2, Gulf, 1974.

Ferziger, J.H., Peric, M., "Computational Methods for Fluid Dynamics", Springer-Verlag, Heidelberg, 1996.

Fewel, K.J., Kean, J.A., "Computer Modeling Aids Separator Retrofit", Oil & Gas Journal, 90(27), 1992, 76-80.

FLOW-3D, Computational Modeling Power for Scientists and Engineers, Flow Science Inc., 1983-2000.

Fluent 6.3.26, "Fluent 6.3 User's Guide", Fluent Inc., Centerra Resource Park, 10 Cavendish Court, Lebanon, USA, 2006.

Fluent 6.3.26, Commercial Computational Fluid Dynamics Simulator, Fluent Inc., 2006.

Frankiewicz, T., Browne, M.M., Lee, C-M., "Reducing Separation Train Sizes and Increasing Capacity by Application of Emerging Technologies", Offshore Technology Conference, OTC-13215, 2001, 1-10.

Frankiewicz, T., Lee, C-M., "Using Computational Fluid Dynamics (CFD) Simulation to Model Fluid Motion in Process Vessels on Fixed and Floating Platforms", Society of Petroleum Engineers, SPE-77494, 2002, 1-9.

Gambit 2.4.6, Preprocessor Tool of Fluent, Fluent Inc., 2006.

Gas Processors Suppliers Association, GPSA Engineering Data Book, 11th Edition, Vol. 1, Gas Processors Association, 1998.

Gerunda, A., "How to Size Liquid-Vapor Separators", Chemical Engineering, May 4, 1981, 81-84.

Gosman, A.D., "Developments in Industrial Computational Fluid Dynamics", Trans IChemE, 76(A), 1998, 153-161.

Green, D.W., Perry, R.H., "Perry's Chemical Engineers' Handbook", 8th Edition, McGraw-Hill, 2008.

Grødal, E.O., Realff, M.J., "Optimal Design of Two- and Three-Phase Separators: A Mathematical Programming Formulation", Society of Petroleum Engineers, SPE 56645, 1999, 1-16.

Hallanger, A., Soenstaboe, F., Knutsen, T., "A Simulation Model for Three-Phase Gravity Separators", Society of Petroleum Engineers, SPE-36644, 1996, 695-706.

Hansen, E.W.M., Celius, H.K., Hafskjold, B., "Fluid Flow and Separation Mechanisms in Offshore Separation Equipment", 1st International Symposium on Two-Phase Flow Modeling and Experimentation, 1995, 117-129.

Hansen, E.W.M., Heitmann, H., Lakså, B., Ellingsen, A., Østby, O., Morrow, T.B., Dodge, F.T., "Fluid Flow Modeling of Gravity Separators", 5th International Conference on Multiphase Production, 1991, 364-380.

Hansen, E.W.M., Heitmann, H., Lakså, B., Løes, M., "Numerical Simulation of Fluid Flow Behavior Inside, and Redesign of a Field Separator", 6th International Conference on Multiphase Production, 1993, 117-129.

Hansen, E.W.M., "Phenomenological Modeling and Simulation of Fluid Flow and Separation Behavior in Offshore Gravity Separators", Emerging Technologies for Fluids, Structures and Fluid-Structure Interaction, 431, 2001, 23-29.

Heidemann, R.A., Jeje, A.A., Mohtadi, F., "An Introduction to the Properties of Fluids and Solids", The University of Calgary Press, 1987.

Helsør, T., Svendsen, H., "Experimental Characterization of Pressure Drop in Dry Demisters at Low and Elevated Pressures", Trans IChemE, 85(A3), 2007, 377-385.

Hesketh, R.P., Fraser Russel, T.W., Etchells, A.W., "Bubble Size in Horizontal Pipelines", AIChE J., 33(4), 1987, 663-667.

Hinze, J.O., "Fundamentals of the Hydrodynamic Mechanism of Splitting in Dispersion Processes", AIChE J., 1(3), 1955, 289-295.

Hooper, W.B., "Decantation", Section 1.11 in "Handbook of Separation Techniques for Chemical Engineers", Ph.A. Schweitzer (Ed.), 3rd Edition, McGraw-Hill, 1997.

HYSYS 3.2, Commercial Chemical Process Simulator, Hyprotech Ltd., 2003.

Ishii, M., Grolmes, M.A., "Inception Criteria for Droplet Entrainment in Two-Phase Concurrent Film Flow", AIChE J., 21(2), 1975, 308-317.

Ishii, M., Zuber, N., "Drag Coefficient and Relative Velocity in Bubbly, Droplet or Paticulate Flows", AIChE J., 25(5), 1979, 843-855.

Issa, R.I., "Solution of the Implicitly Discretized Fluid Flow Equations by Operator-Splitting", Journal of Computational Physics, 62, 1986, 40-65.

Karabelas, A.J., "Droplet Size Spectra Generated in Turbulent Pipe Flow of Dilute Liquid/Liquid Dispersions", AIChE J., 24(2), 1978, 170-180.

Kim, H., Burgess, D.J., "Prediction of Interfacial Tension between Oil Mixtures and Water", J. Collide Interface Sci., 241, 2001, 509-513.

King, R.P., "Introduction to Practical Fluid Flow", Butterworth-Heinemann, 2002.

Kolmogorov, A.N., "On the Breaking of Drops in Turbulent Flow", Doklady Akad. Nauk. SSSR, 66, 1949, 825-828.

Kolodzie, P.A., Van Winkle, M., "Discharge Coefficients through Perforated Plates", AIChE J., 3(3), 1957, 305-312.

Kumar, A., Hartland, S., "Gravity Settling in Liquid/Liquid Dispersions", Can. J. Chem. Eng., 63(3), 1985, 368-276.

Launder, B.E., Spalding, D.B., "Mathematical Models of Turbulence", Academic Press, 1972.

Lee, C-M., Dijk, E.V., Legg, M., "Field Confirmation of CFD Design for FPSO-mounted Separator", Offshore Technology Conference, OTC-16137, 2004, 1-6.

Lee, J.M., Khan, R.I., Phelps, D.W., "Debottlenecking and Computational Fluid Dynamics Studies of High and Low-Pressure Production Separators", Society of Petroleum Engineers, SPE-115735, 2009, 124-131.

Levich, V.G., "Physiochemical Hydrodynamics", Prentice Hall, 1962.

Lu, Y., Lee, J.M., Phelps, D., Chase, R., "Effect of Internal Baffles on Volumetric Utilization of a FWKO - A CFD Evaluation", Society of Petroleum Engineers, SPE-109944, 2007, 1-6.

Lyons, W.C., Plisga, G.J. (Editors), "Standard Handbook of Petroleum and Natural Gas Engineering", Volume 2, Gulf Professional Publishing, 2005.

McCain, W.D., "The Properties of Petroleum Fluids", Petroleum Publishing Company, 1973.

Monnery, W.D., Svrcek, W.Y., "Successfully Specify Three-Phase Separators", Chem. Eng. Progress, 90(9), 1994, 29-40.

Monnery, W.D., Svrcek, W.Y., "Analytical Study of Liquid/Vapor Separation Efficiency", Technical Research Report, Petroleum Technology Alliance Canada, 2000.

Navier, C.L.M.H., "Mémoire sur les lois du movement des fluides", Mémoires de VAcadémie des Sciences de VInstitut de France , 6, 1822, 389-440.

Newton, T., Connolly, D., Mokhatab, S., "Tools to Model Multiphase Separation", Chem. Eng. Progress, 103(6), 2007, 26-31.

Patankar, S.V., Spalding, D.B., "A Calculation Procedure for Heat, Mass and Momentum Transfer in Three-dimensional Parabolic Flows", International Journal of Heat and Mass Transfer, 15, 1972, 1787–1806.

Perry, R.H., Green, D.W., Maloney, J.O., "Perry's Chemical Engineers' Handbook", 7th Edition, McGraw-Hill, 1999.

Poling, B.E., Prausnitz, J.M., O'Connell, J.P., "The Properties of Gases and Liquids", 5th Edition, McGraw-Hill, 2001.

Rosin, P., Rammler, E., "The Laws Governing the Fineness of Powdered Coal", J. Inst. Fuel, 7, 1933, 29-36.

Sharratt, P.N., "Computational Fluid Dynamics and its Application in the Process Industries", Trans IChemE, 68-A, January, 1990, 13-18.

Shelley, S., "Computational Fluid Dynamics – Power to the People", Chem. Eng. Progress, 103(4), April, 2007, 10-13.

Sinnott, R.K., "Chemical Engineering Design" in "Coulson & Richardson's Chemical Engineering", 2nd Edition, Butterworth-Heinemann, 1997.

Skelton, G.F., "Production", in "Our Industry Petroleum", Stockil, P.A. (Ed.), British Petroleum Company Limited, 1977.

Sleicher, C.A., "Maximum Stable Drop Size in Turbulent Flow", AIChE J., 8(4), 1962, 471-477.

Smith, H.V., "Oil and Gas Separators", in "Petroleum Engineering Handbook", Bradley, H.B. (Ed), Society of Petroleum Engineers, 1987.

Stokes, G.G., "On the Theories of Internal Friction of Fluids in Motion, and of the Equilibrium and Motion of Elastic Solids", Transaction of the Cambridge Philosophical Society, 8(22), 1845, 287-305.

Streeter, V.L., Wylie, E.B., "Fluid Mechanics", 8th Edition, McGraw-Hill, 1985.

Svrcek, W.Y., Monnery, W.D., "Design Two-Phase Separators within the Right Limits", Chem. Eng. Progress, 89(10), 1993, 53-60.

Swartzendruber, J., Fadda, D., Taylor, D., "Accommodating Last Minute Changes: Two Phase Separation Performance Validated by CFD", ASME Fluids Engineering Division Summer Meeting and Exhibition, Proceedings of FEDSM, 2005, 713-715.

Tatterson, D.F., Dallman, J.C., Hanratty, T.J., "Drop Sizes in Annular Gas-Liquid Flows", AIChE J., 23(1), 1977, 68-76.

Verlaan, C.C.J., Olujic, Z., De Graauw, J., "Performance Evaluation of Impingement Gas-Liquid Separators", Proceedings of the 4th International Conference on Multiphase Flow, 1989, 103-115.

Vetter, O.J., Bent, M., Kandarpa, V., Salzman, D., Williams, R., "Three-Phase PVT and CO_2 Partitioning", Society of Petroleum Engineers, SPE 16351, 1987, 297-310.

Viles, J.C., "Predicting Liquid Re-Entrainment in Horizontal Separators", Society of Petroleum Engineers, Journal of Petroleum Technology, 1993, 405-409.

Visser, R.C., "Offshore Production of Heavy Oil", J. Petroleum Technology, 1989, 67-70.

Walas, S.M., "Process Vessels", Chapter 18 in "Chemical Process Equipment Selection and Design", Butterworth-Heinemann, 1990.

Watkins, R.N., "Sizing Separators and Accumulators", Hydrocarbon Proc., 46(11), 1967.

Wilkinson, D., Waldie, B., "CFD and Experimental Studies of Fluid and Particle Flow in Horizontal Primary Separators", Trans IChemE, 72(A), 1994, 189-196.

Wilkinson, D., Waldie, B., Nor, M.I.M., Lee, H.Y., "Baffle Plate Configuration to Enhance Separation in Horizontal Primary Separators", Chem. Eng. J., 77, 2000, 221-226.

Wu, F.H., "Separators, Liquid-Vapor, Drum Design", in "Encyclopedia of Chemical Processing and Design", J.J. McKetta, W.A. Cunningham (Ed.), Marcel Dekker, 1990.

Appendix A: Design of Three-Phase Separators

The aim of Appendix A is to provide the approach and the main steps proposed by Monnery and Svrcek (1994) for designing the optimum three-phase separator. In their design approach, the economical aspect ratio of a separator is assumed to be between 1.5 and 6 with a functionality of operating pressure as presented in Table A-1.

A.1 Vertical Configuration

First, the vessel diameter is calculated based on satisfactory separation of oil droplets from gas phase. Then, the heights of the light and heavy liquids are assumed, and the liquid-liquid separation velocities and times are calculated. If the residence times of the continuous liquid phases are not larger than the required separation times, then the vessel diameter will be increased to satisfy the liquid-liquid separation requirements. Figure A-1 shows a schematic of the vertical three-phase separator for which the design procedure is outlined here:

1. Calculate the terminal settling velocity of oil droplets using Equation 2-1 or Equation 2-4.
2. Set the vapor velocity equal to 75% of the terminal settling velocity.
3. Calculate all of the volumetric flow-rates.
4. Calculate the vessel diameter based on the vapor volumetric flow-rate and the vapor velocity.
5. Calculate the separation velocities of both the liquid phases through each other using Stokes' law.
6. Set the thickness of liquid phases (assume $H_L = H_H = 30\ cm$ as minimum), and calculate the separation times for the liquid droplets (t_{HL} and t_{LH}).
7. Calculate the cross-sectional areas of the liquid phases. Note, this area is the same as the vessel cross-sectional area for heavy liquid phase ($A_H = A$), but in the case of using a baffle plate down-comer, the area allotted to baffle plate should be subtracted from the vessel cross-sectional area to obtain the area of light liquid phase ($A_L = A - A_D$).
8. Calculate the residence time of the light liquid: $\theta_L = \dfrac{H_L A_L}{Q_{LL}}$. If $\theta_L < t_{HL}$, increase the vessel diameter so that $\theta_L = t_{HL}$.

Table A-1. The aspect ratio values suggested by Monnery and Svrcek (1994) for multiphase separators.

P (kPa)	0-1700	1700-3400	0>3400
Separator Aspect Ratio	1.5-3	3-4	4-6

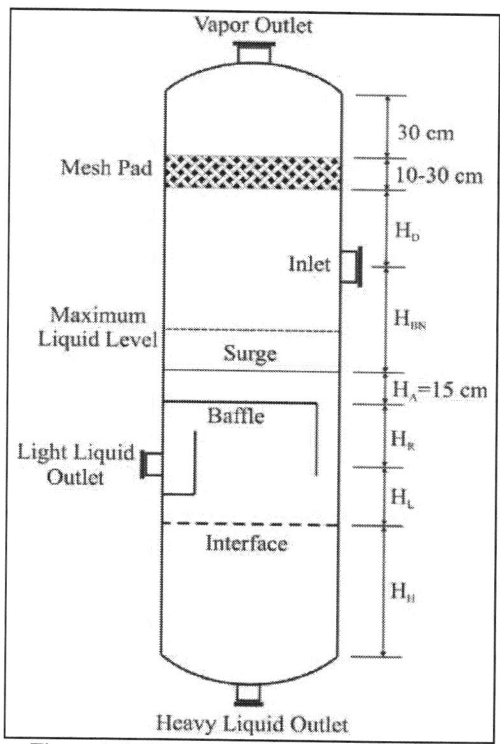

Figure A-1. The Vertical Three-Phase Separator.

9. Calculate the residence time of the heavy liquid: $\theta_H = \frac{H_H A_H}{Q_{HL}}$. If $\theta_H < t_{LH}$, increase the vessel diameter so that $\theta_H = t_{LH}$.
10. Calculate the height of light liquid phase above the outlet (H_R) using the holdup time.

11. Calculate the surge height (H_S) based on the surge time.

12. Calculate the vessel height (H_T). If $\frac{H_T}{D}$ is not in the range of 1.5-6.0, increase the diameter (to decrease the ratio) or height (to increase the ratio) of the separator and fix the problem.

A.2 Horizontal Configuration

An iterative design procedure is required to determine the most economical separator. At each iteration, with an assumed diameter, the vapor disengagement area is set to provide a satisfactory separation of liquid droplets from gas phase, and the heights of the light and heavy liquids are assumed. Then, the lengths required by vapor-liquid separation and retention time requirements are calculated. Similar to the vertical configuration, if the residence times are not greater than the required separation times, then the separator size should be increased. Note, when increasing the length (preferably) or diameter of a separator, the separator aspect ratio should be in the acceptable range of 1.5-6.

The different common designs of horizontal three-phase separator, composed of "simple", "boot", "weir", and "bucket and weir", are illustrated in Figure A-2 to Figure A-5, respectively. In design procedure, first, the terminal settling velocity of oil droplets is estimated using Equation 2-1 or Equation 2-4 and vapor velocity is set to be equal to 75% of the terminal settling velocity. Then, holdup and surge volumes (V_H and V_S) are calculated based on given holdup and surge times and volumetric flow-rate of light liquid (Q_{LL}). The other steps differ from one design to another and are outlined in the following subsections.

A.2.1 "Simple" Design

1. Pick an aspect ratio from Table A-1 and calculate the initial vessel diameter: $D = \sqrt[3]{\frac{4(V_H + V_S)}{0.5\pi\left(\frac{L}{D}\right)}}$.

2. Calculate vessel cross-sectional area: $A = \frac{\pi D^2}{4}$.

3. Set H_H and H_L (assume $H_L = H_H = 30\ cm$ as minimum), and calculate $(A_H + A_L)$.
4. Set H_V to the larger of $0.20 \times D$ or 0.60 *m*, and then calculate A_V.
5. Calculate the vessel length based on the liquid holdup/surge: $L = \frac{V_H + V_S}{A - A_V - (A_H + A_L)}$.
6. Calculate the liquid dropout time: $\phi = \frac{H_V}{V_V}$.
7. Calculate the actual vapor velocity using Q_V and A_V.
8. Calculate the minimum length required for vapor-liquid separation (L_{min}) using the actual vapor velocity and the liquid dropout time.
9. If $L < 0.8L_{\min}$, increase H_V, and go to step 4. Else if $L < L_{\min}$, set $L = L_{\min}$. Else if $L > 1.2L_{\min}$, decrease H_V (if acceptable), and go to step 4. Else, L is acceptable.
10. Calculate the separation velocities of both the liquid phases through each other using Stokes' law.
11. Calculate the separation times of the liquid droplets (t_{HL} and t_{LH}).
12. Calculate the residence time of the light liquid: $\theta_L = \frac{L(A - A_V - A_H)}{Q_{LL}}$. If $\theta_L < t_{HL}$, set

$$L = \frac{t_{HL} Q_{LL}}{A - A_V - A_H}.$$

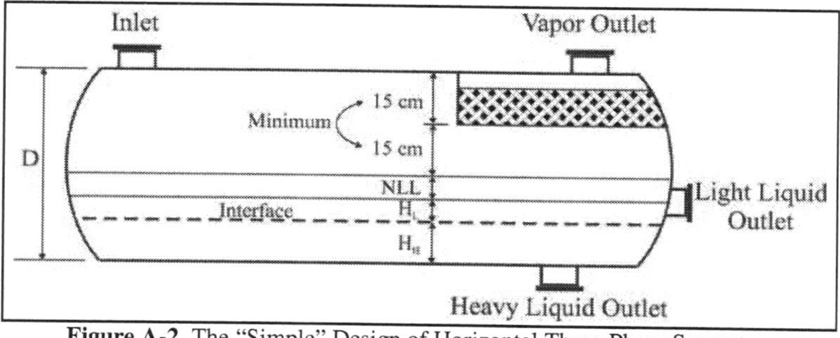

Figure A-2. The "Simple" Design of Horizontal Three-Phase Separator.

13. Calculate the residence time of the heavy liquid: $\theta_H = \frac{LA_H}{Q_{HL}}$. If $\theta_H < t_{LH}$, set $L = \frac{t_{LH} Q_{HL}}{A_H}$.
14. If $\frac{L}{D} < 1.5$, decrease D (if acceptable), and go to step 2. Else if $\frac{L}{D} > 6$, increase D, and go to step 2.
15. Calculate the approximate vessel weight based on thickness and surface area of shell and heads.
16. In order to find the optimum case (corresponding to the minimum weight), change the vessel diameter by 15 cm increments, and repeat the calculations from step 2 while keeping the aspect ratio in the range of 1.5 to 6.0.

A.2.2 "Boot" Design

1. Pick an aspect ratio from Table A-1 and calculate the initial vessel diameter: $D = \sqrt[3]{\frac{4(V_H + V_S)}{0.5\pi\left(\frac{L}{D}\right)}}$.
2. Calculate vessel cross-sectional area: $A = \frac{\pi D^2}{4}$.
3. Set H_V to the larger of $0.20 \times D$ or 0.60 m, and then calculate A_V.
4. Set H_{LLV} and H_{LLB}, and then calculate A_{LLV}.
5. Calculate the vessel length based on the liquid holdup/surge: $L = \frac{V_H + V_S}{A - A_V - A_{LLV}}$.
6. Calculate the liquid dropout time: $\phi = \frac{H_V}{V_V}$.
7. Calculate the actual vapor velocity using Q_V and A_V.
8. Calculate the minimum length required for vapor-liquid separation (L_{min}) using the actual vapor velocity and the liquid dropout time.
9. If $L < 0.8L_{\min}$, increase H_V, and go to step 3. Else if $L < L_{\min}$, set $L = L_{\min}$. Else if $L > 1.2L_{\min}$, decrease H_V (if acceptable), and go to step 3. Else, L is acceptable.

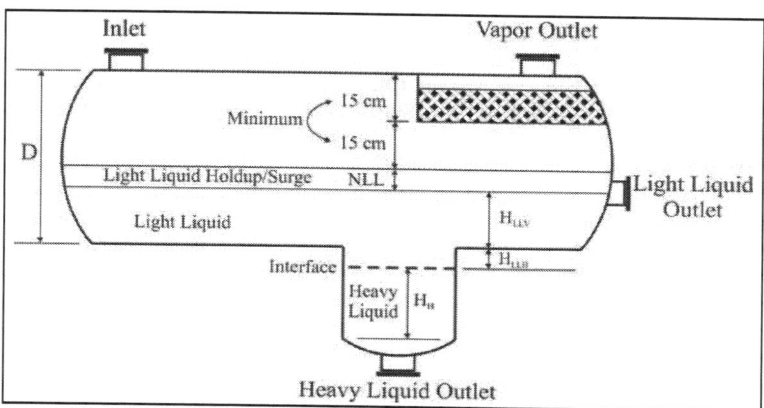

Figure A-3. The "Boot" Design of Horizontal Three-Phase Separator.

10. Calculate the settling velocity of the heavy liquid through the light liquid using Stokes' law.
11. Calculate the settling time of the heavy liquid through the light liquid (t_{HL}).
12. Calculate the residence time of the light liquid: $\theta_L = \frac{L(A - A_V)}{Q_{LL}}$. If $\theta_L < t_{HL}$, set $L = \frac{t_{HL} Q_{LL}}{A - A_V}$.
13. If $\frac{L}{D} < 1.5$, decrease D (if acceptable), and go to step 2. Else if $\frac{L}{D} > 6$, increase D, and go to step 2.
14. Calculate the approximate vessel weight based on the thickness and the surface area of shell and heads.
15. In order to find the optimum case (corresponding to the minimum weight), change the vessel diameter by 15 cm increments, and repeat the calculations from step 2 while keeping the aspect ratio in the range of 1.5 to 6.0.
16. Design the heavy liquid boot:

    16.1. Set H_H.

16.2. Calculate the rising velocity of the light liquid out of the heavy liquid using Stokes' law, and use 75% of this velocity as V_B in the calculations.

16.3. Calculate the heavy liquid boot diameter, D_B, using Q_{HL} and V_B.

16.4. Calculate the rising time of the light liquid droplets through the heavy liquid (t_{LH}).

16.5. Calculate the residence time of the heavy liquid: $\theta_H = \dfrac{\pi D_B^{\,2} H_H}{4Q_{HL}}$. If $\theta_H < t_{LH}$, increase the boot diameter so that $\theta_H = t_{LH}$.

A.2.3 "Weir" Design

1. Pick an aspect ratio from Table A-1 and calculate the initial vessel diameter: $D = \sqrt[3]{\dfrac{16(V_H + V_S)}{0.6\pi\left(\dfrac{L}{D}\right)}}$.
2. Calculate vessel cross-sectional area: $A = \dfrac{\pi D^2}{4}$.
3. Set H_V to the larger of $0.20 \times D$ or 0.60 m, and then calculate A_V.
4. Set the low liquid level (H_{LLL}) in light liquid section, and then calculate A_{LLL}.
5. Calculate the weir height: $H_W = D - H_V$. If $H_W < 60\ cm$, increase D, and go to step 2.
6. Calculate L_2 based on the light liquid holdup/surge: $L_2 = \dfrac{V_H + V_S}{A - A_V - A_{LLL}}$.
7. Set the interface at $\dfrac{H_W}{2}$ (typical setting), and obtain H_H and H_L.
8. Using H_H value calculate A_H, and set $A_L = A - A_V - A_H$.
9. Calculate the separation velocities of liquid phases using Stokes' law.
10. Calculate the separation times of the liquid droplets (t_{HL} and t_{LH}).
11. Set the larger of $\dfrac{t_{LH} Q_{HL}}{A_H}$ and $\dfrac{t_{HL} Q_{LL}}{A_L}$ as the required length for liquid-liquid separation (L_1).
12. Set $L = L_1 + L_2$.

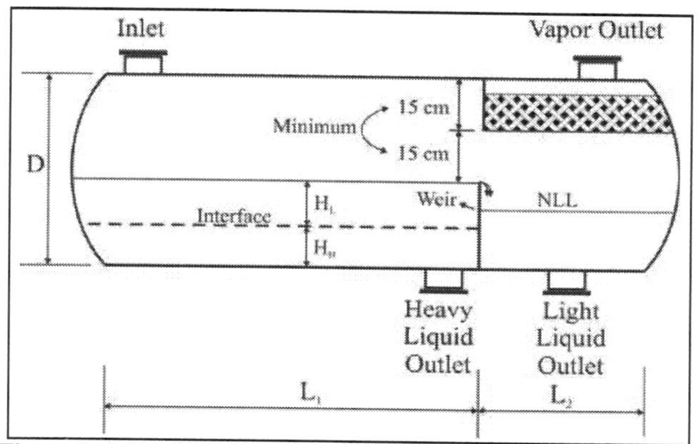

Figure A-4. The "Weir" Design of Horizontal Three-Phase Separator.

13. Calculate the liquid dropout time: $\phi = \dfrac{H_V}{V_V}$.
14. Calculate the actual vapor velocity using Q_V and A_V.
15. Calculate the minimum length required for vapor-liquid separation (L_{min}) using the actual vapor velocity and the liquid dropout time.
16. If $L < 0.8L_{\min}$, increase H_V, and go to step 3. Else if $L < L_{\min}$, set $L = L_{\min}$. Else if $L > 1.2L_{\min}$, decrease H_V (if acceptable), and go to step 3. Else, L is acceptable.
17. If $\dfrac{L}{D} < 1.5$, decrease D (if acceptable), and go to step 2. Else if $\dfrac{L}{D} > 6$, increase D, and go to step 2.
18. Calculate the approximate vessel weight based on the thickness and the surface area of shell and heads.
19. In order to find the optimum case (corresponding to the minimum weight), change the vessel diameter by 15 cm increments, and repeat the calculations from step 2 while keeping the aspect ratio in the range of 1.5 to 6.0.

A.2.4 "Bucket and Weir" Design

1. Assume residence times of light and heavy liquid phases, θ_L and θ_H.
2. Pick an aspect ratio from Table A-1 and calculate the initial vessel diameter: $D = \sqrt[3]{\dfrac{4(Q_{LL}\theta_L + Q_{HL}\theta_H)}{0.7\pi\left(\dfrac{L}{D}\right)}}$.
3. Calculate vessel cross-sectional area: $A = \dfrac{\pi D^2}{4}$.
4. Set H_V to the larger of $0.20 \times D$ or 0.60 *m*, and then calculate A_V.
5. Calculate L_1: $L_1 = \dfrac{Q_{LL}\theta_L + Q_{HL}\theta_H}{A - A_V}$.
6. Calculate the liquid dropout time: $\phi = \dfrac{H_V}{V_V}$.
7. Calculate the actual vapor velocity using Q_V and A_V.
8. Calculate the minimum length required for vapor-liquid separation (L_{min}) using the actual vapor velocity and the liquid dropout time.
9. If $L_1 < 0.8L_{\min}$, increase H_V, and go to step 4. Else if $L_1 < L_{\min}$, set $L_1 = L_{\min}$. Else, L_1 is acceptable.
10. Calculate H_L using Stokes' law and θ_L.
11. Calculate the height difference between the light and heavy liquid weirs: $\Delta H = H_L\left(1 - \dfrac{\rho_L}{\rho_H}\right)$.
12. Design the light liquid bucket:
    - 12.1. Set the top of the light liquid weir.
    - 12.2. Assume holdup and surge times.
    - 12.3. Assume H_{HLL}, and calculate A_{HLL}.
    - 12.4. Assume H_{LLL}, and calculate A_{LLL}.
    - 12.5. Calculate L_2: $L_2 = \dfrac{Q_{LL}(t_H + t_S)}{A_{HLL} - A_{LLL}}$.

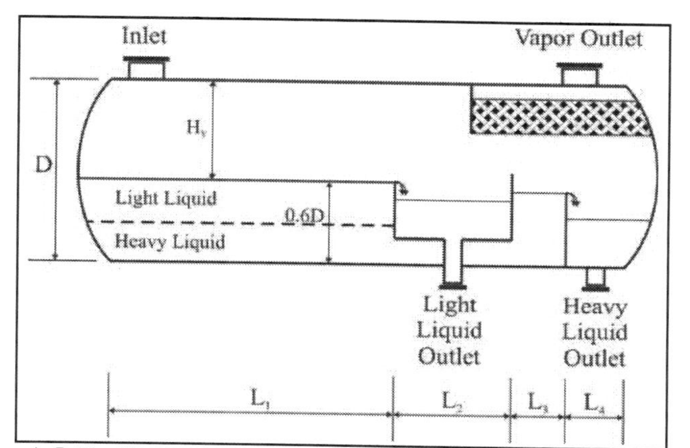

Figure A-5. The "Bucket and Weir" Design of Horizontal Three-Phase Separator.

13. Set L_3 as the larger of $D/12$ or 30 cm.

14. Design the heavy liquid section:

    14.1. Set the top of the heavy liquid weir.

    14.2. Assume holdup and surge times.

    14.3. Assume H_{HLL}, and calculate A_{HLL}.

    14.4. Assume H_{LLL}, and calculate A_{LLL}.

    14.5. Calculate L_4: $L_4 = \dfrac{Q_{HL}(t_H + t_S)}{A_{HLL} - A_{LLL}}$.

15. Set $L = L_1 + L_2 + L_3 + L_4$.

16. If $\dfrac{L}{D} < 1.5$, decrease D (if acceptable), and go to step 3. Else if $\dfrac{L}{D} > 6$, increase D, and go to step 3.

17. Calculate the approximate vessel weight based on thickness and surface area of shell and heads.

18. In order to find the optimum case (corresponding to the minimum weight), change the vessel diameter by 15 cm increments, and repeat the calculations from step 3 while keeping the aspect ratio in the range of 1.5 to 6.0.

Appendix B: Droplet Breakups in the Two-Phase Separators

In this Appendix, the diagrams of droplet breakup variations versus continuous gas phase velocity for all the pilot-plant scale separators have been presented. As explained in Chapter Three, two modeling approaches were implemented: DPM approach and VOF-DPM approach.

B.1 Results for DPM Approach

Figures B-1 to B-4 show the obtained diagrams for the number of droplet breakups versus gas phase velocity in all the case studies. As represented, three different operating pressures were involved, and three different mean droplet sizes were also tested in the simulations. The presented variations indicate that higher velocities and operating pressures usually intensify the number of droplet breakups. However, the horizontal separators usually proposed a maximum breakup point so that before this point, which was independent of the incipient velocity, increasing the velocity led to an increase in the droplet breakup number, and after this maximum point, increasing the velocity decreased the number of breakups. The simulation results also showed that changing the mean droplet size when keeping the other size distribution parameters constant, has no tangible effect on the droplet breakup number.

B.2 Results for VOF-DPM Approach

Figures B-5 to B-8 present the obtained diagrams for number of droplet breakups versus gas phase velocity in all the case studies. The trend of variations in the horizontal separators indicates that higher velocities usually intensified the number of droplet breakups, and higher pressures had a stabilizing effect on the droplet breakup variations. Moreover, increasing the separator aspect ratio was partially in favor of droplet breakups.

In vertical separators, droplet breakup variations versus gas velocity tended to be somewhat oscillating at low pressures, but at higher pressures the number of breakups was stabilized to a constant rate of about 20%.

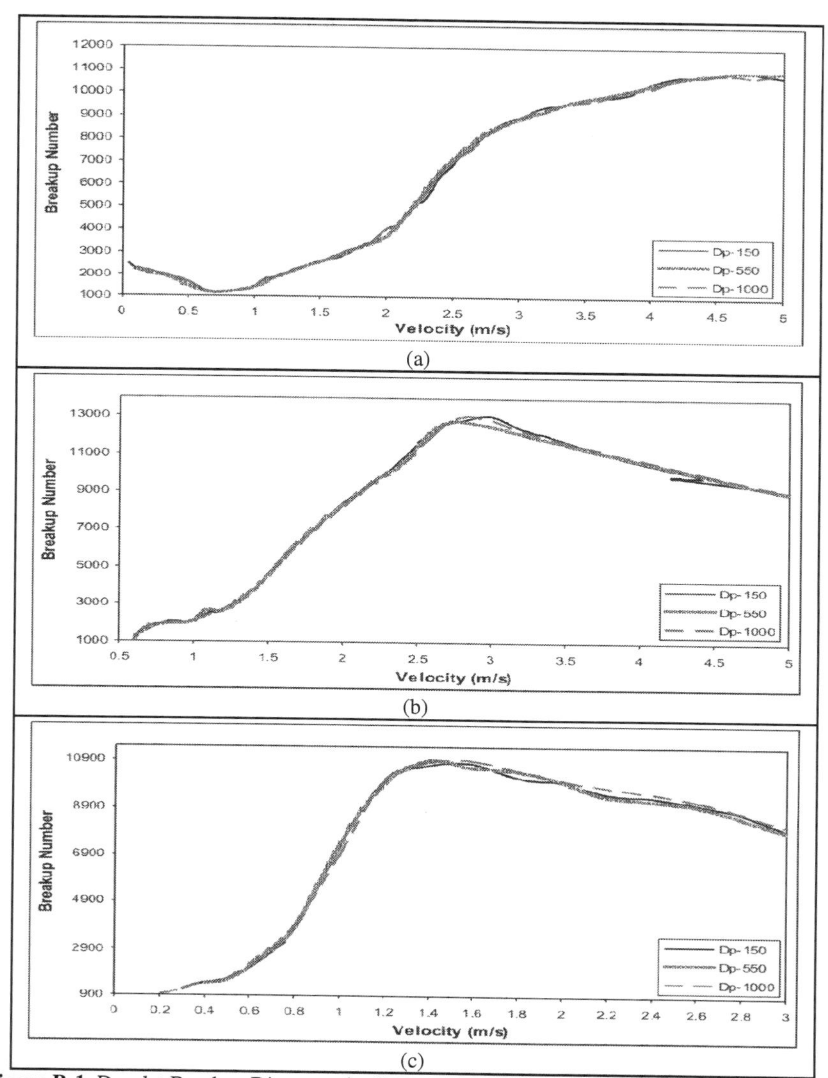

Figure B-1. Droplet Breakup Diagrams for Separator A at Multiple Pressures (DPM Approach); (a) 70 *kPa*, (b) 700 *kPa*, and (c) 2760 *kPa*.

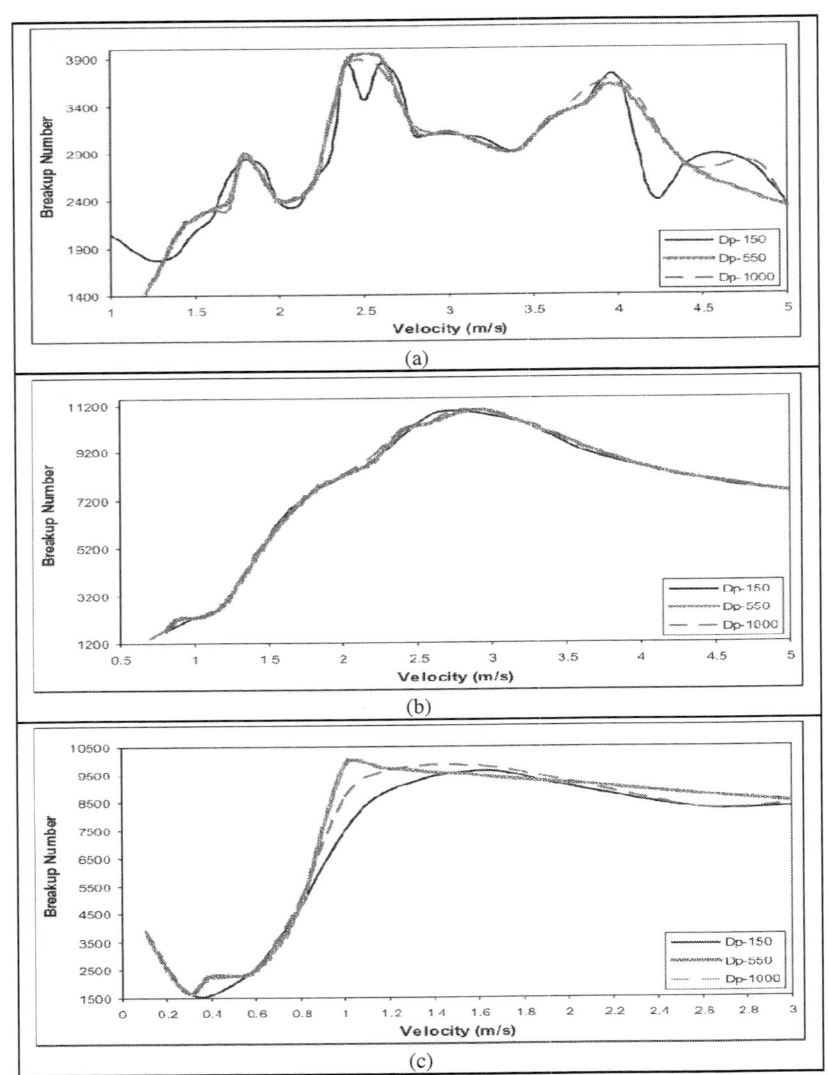

Figure B-2. Droplet Breakup Diagrams for Separator B at Multiple Pressures (DPM Approach); (a) 70 *kPa*, (b) 700 *kPa*, and (c) 2760 *kPa*.

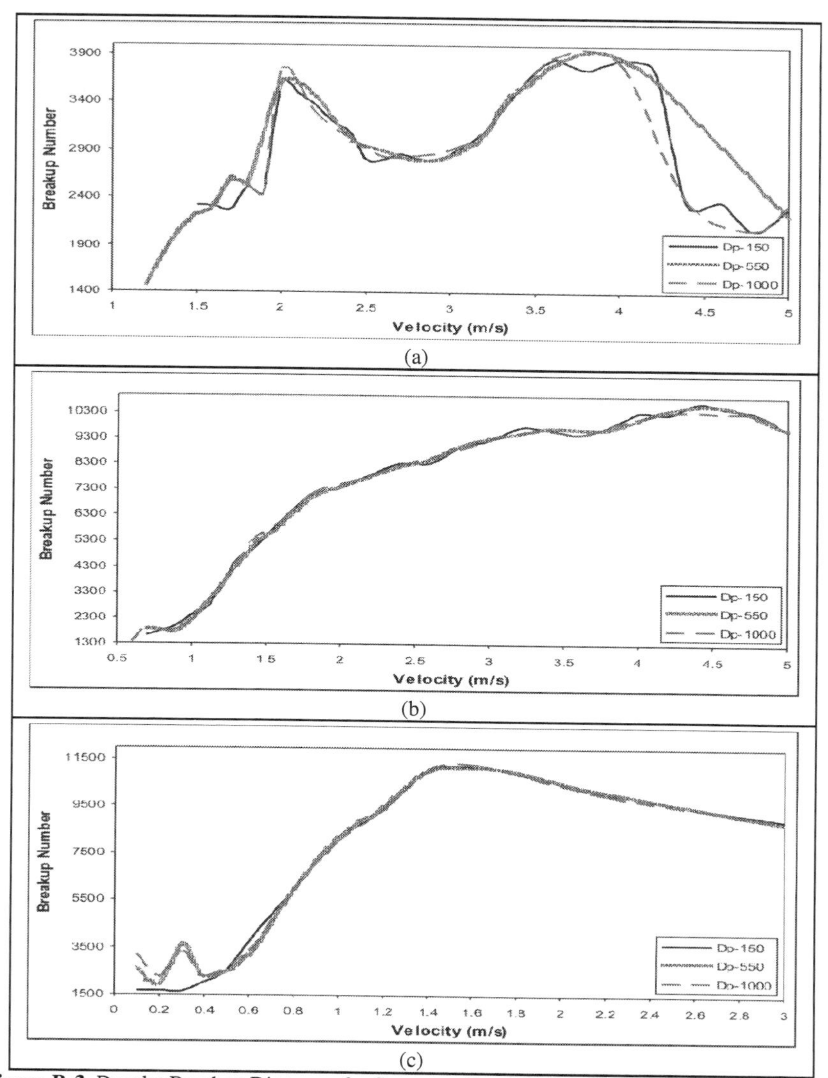

Figure B-3. Droplet Breakup Diagrams for Separator C at Multiple Pressures (DPM Approach); (a) 70 *kPa*, (b) 700 *kPa*, and (c) 2760 *kPa*.

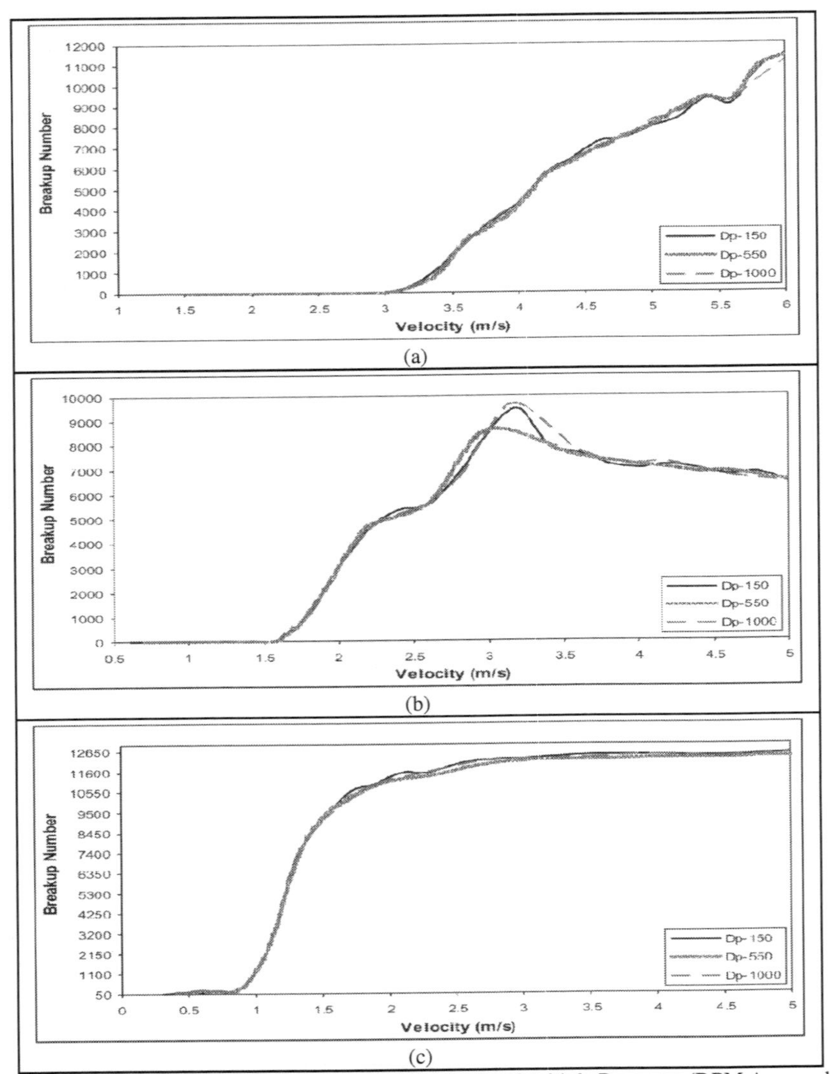

Figure B-4. Droplet Breakup Diagrams for Separator D at Multiple Pressures (DPM Approach); (a) 70 *kPa*, (b) 700 *kPa*, and (c) 2760 *kPa*.

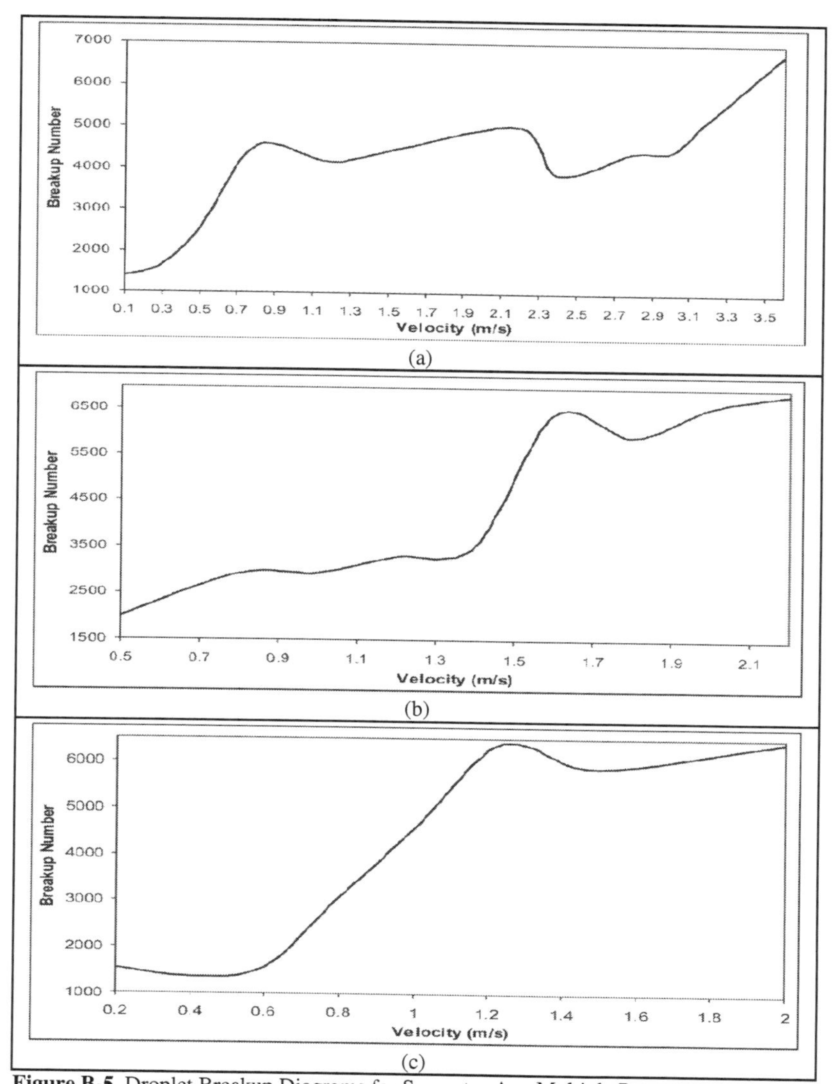

Figure B-5. Droplet Breakup Diagrams for Separator A at Multiple Pressures (VOF-DPM Approach); (a) 70 *kPa*, (b) 700 *kPa*, and (c) 2760 *kPa*.

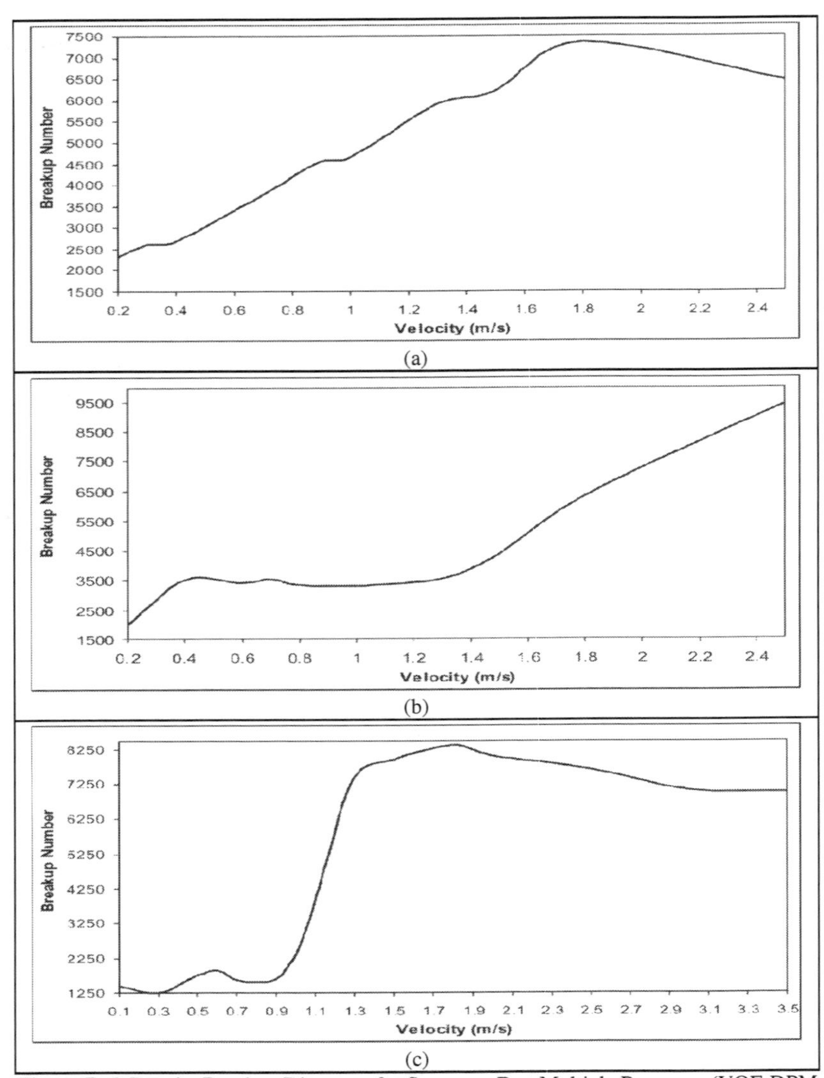

Figure B-6. Droplet Breakup Diagrams for Separator B at Multiple Pressures (VOF-DPM Approach); (a) 70 *kPa*, (b) 700 *kPa*, and (c) 2760 *kPa*.

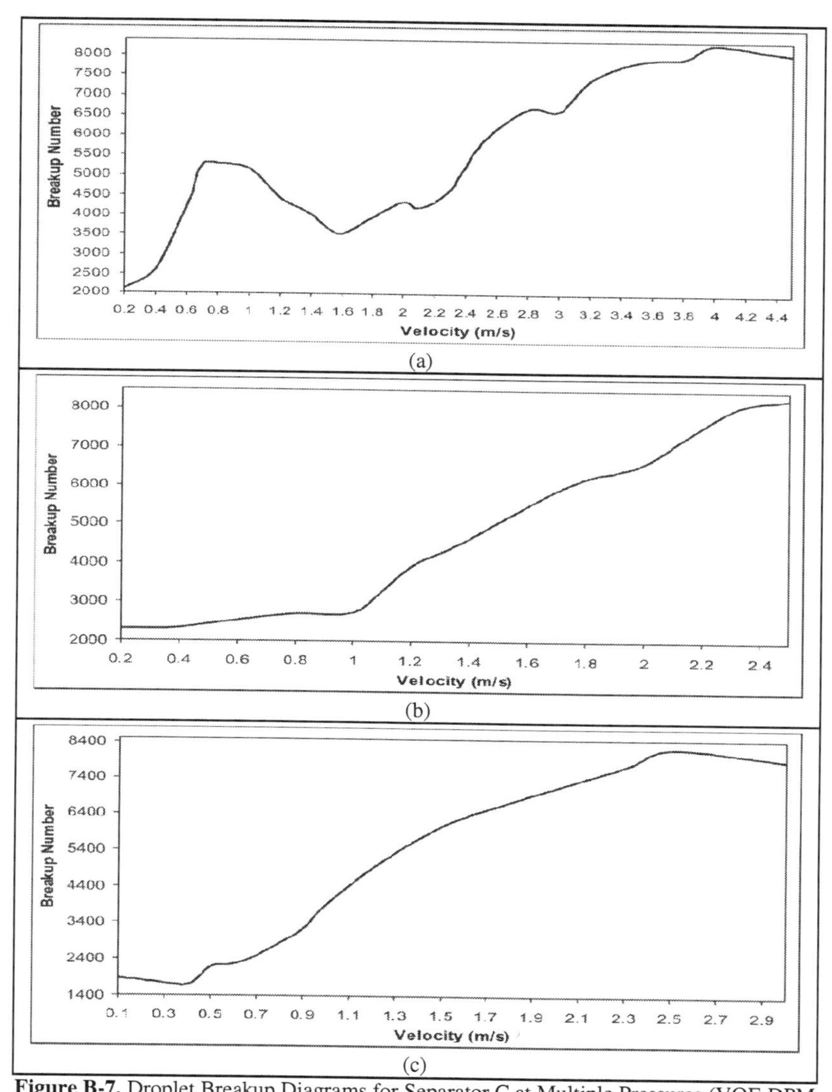

Figure B-7. Droplet Breakup Diagrams for Separator C at Multiple Pressures (VOF-DPM Approach); (a) 70 *kPa*, (b) 700 *kPa*, and (c) 2760 *kPa*.

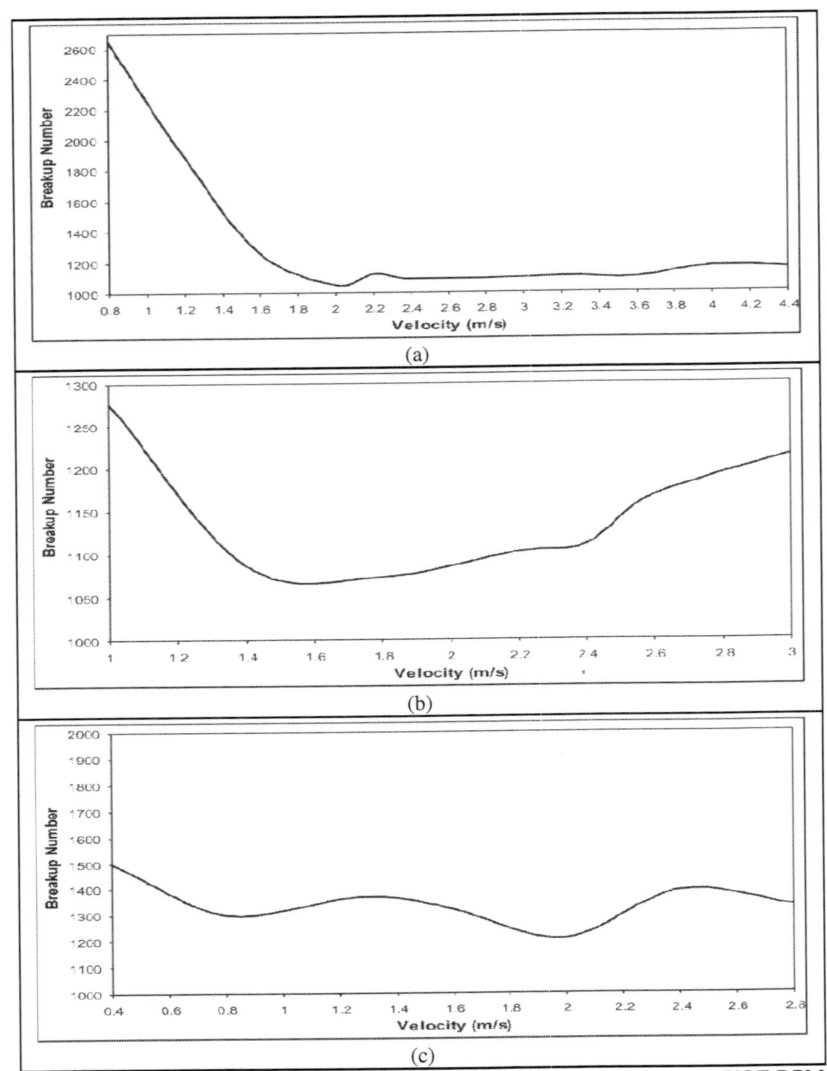

Figure B-8. Droplet Breakup Diagrams for Separator D at Multiple Pressures (VOF-DPM Approach); (a) 70 *kPa*, (b) 700 *kPa*, and (c) 2760 *kPa*.

Appendix C: Separation Performance of Wire Mesh Demisters

As explained in Chapter Three, the demister of Gullfaks-A separator was modeled using the Fluent Porous Media Model. The results of CFD simulations, presented in Chapter Four, indicated that pressure and velocity profiles assigned to the demister were correct. However, since almost all the liquid droplets had already settled from gas stream, separation performance of the demister could not be verified. To examine this issue, a CFD model was developed for the configuration represented in Figure C-1. This typical demister is 0.15 *m* thick with physical properties as reported by Helsør and Svendsen (2007), Table 3-10. Chapter Three provides more details of the developed CFD model for the wire mesh demister. The oil, water, and gas properties of the Gullfaks-A case study were used. Two particle size distributions were defined (Table C-1) to study the effect of size distribution on the demister separation efficiency.

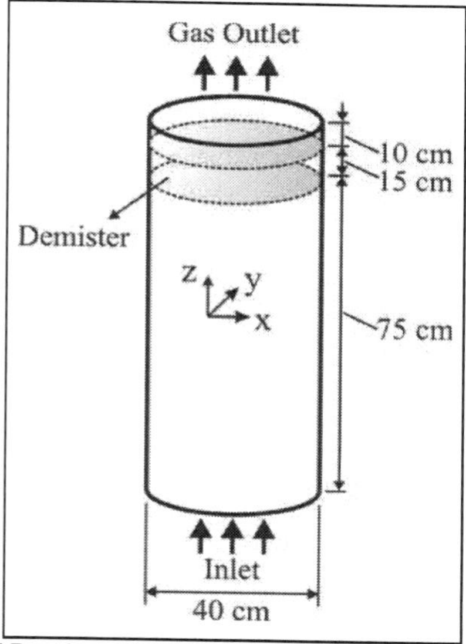

Figure C-1. The Model Developed for CFD Simulation of Demister Separation Efficiency.

Table C-1. The discrete phase parameters used in CFD simulations of demister performance.

Discrete Phase Parameters	Expected for Demister Vicinity		Extended Size Distribution	
	Oil Drops	Water Drops	Oil Drops	Water Drops
Maximum Diameter (μm)	150	150	2000	2000
Mean Diameter (μm)	60	60	800	800
Total Mass Flow-Rate (kg/s)	1.9×10^{-7}	2.3×10^{-7}	4.5×10^{-4}	5.5×10^{-3}
Number of Tracked Particles	1000			
Minimum Diameter (μm)	5			
Spread Parameter	2.6			

The original particle size distribution was assumed based on the fact that the main role of demisters is to remove very fine droplets, 10 to 100 μm, from the gas stream (Smith, 1987). Note, while the maximum droplet size was increased from 150 μm to 2000 μm for the case of extended droplet size distribution, the minimum droplet size was set at a constant value of 5 μm for both distributions.

In the developed model, the gas velocity was gradually increased from 0.02 to 4.0 *m/s* and the demister separation efficiency was calculated using Equation C-1:

$$\eta_{demister} = \frac{m_{in} - m_{out}}{m_{in}} \times 100 \qquad \textbf{(C-1)}$$

where m_{in} and m_{out} are mass of entered to and exited from the demister in *kg*, respectively.

Figures C-2 and C-3 show the separation efficiency of demister for the two distributions of oil and water droplets as a function of gas velocity. These figures do show that the separation efficiency of demister has a maximum value at a very low velocity and then drops sharply to very low values. Therefore, as was expected from experience (Coker, 2007), the increase in gas velocity within the operating range has resulted in an improved demister performance, until a maximum safe velocity is reached, after which increasing gas velocity causes rapid loss of demister efficiency.

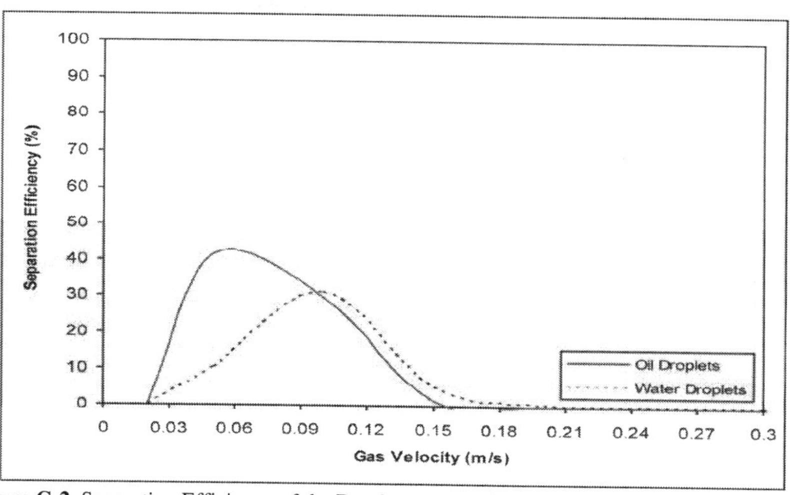

Figure C-2. Separation Efficiency of the Demister versus Gas Velocity for the Original Droplet Size Distribution.

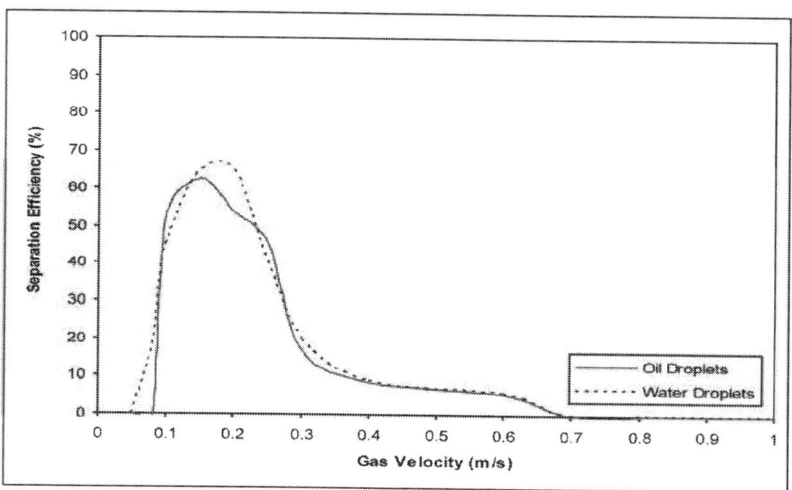

Figure C-3. Separation Efficiency of the Demister versus Gas Velocity for the Extended Droplet Size Distribution.

Both the maximum demister efficiency and the corresponding gas velocity are higher for the case of the extended droplet size distribution, which is consistent with the practical experience in that large droplets are expected to be separated more efficiently than smaller droplets. However, from a practical perspective, the maximum demister efficiency is expected to be close to 100% for droplets with a diameter of 100 μm or less (down to around 5 μm), and this performance should occur for the higher gas velocities when compared with the simulated results. Therefore, it can be concluded that although the general behavior of mist eliminators has been simulated, the maximum demister efficiency and the range of demister operability were predicted much more conservatively.

Another CFD-based study of the demister performance was provided by the droplet size distribution analysis in inlet and outlet zones. Table C-2 reports the droplet size distribution in inlet and outlet zones of the demister operating at its best performance. If the results given for the demister outlet zone are compared with those given for the demister inlet zone, it is evident that the demister had no problem in separating the larger droplets and reducing the maximum and mean droplet sizes. Moreover, the droplet size distribution in the demister outlet zone, with higher values assigned for the spread parameter, is narrower than that in the demister inlet zone.

Table C-2. The droplet size distribution in inlet and outlet zones of the demister.

Discrete Phase Parameters		Original Size Distribution		Extended Size Distribution	
		Demister Inlet	Demister Outlet	Demister Inlet	Demister Outlet
Oil Drops	d_{min} (μm)	5	5	5	5
	d_{max} (μm)	51	48	129	99
	$\bar{d}$ (μm)	48	35	152	96
	n	3.55	3.72	2.65	2.69
Water Drops	d_{min} (μm)	5	5	5	5
	d_{max} (μm)	61	57	167	161
	$\bar{d}$ (μm)	51	41	234	129
	n	3.49	3.62	2.57	2.64

To further investigate the demister separation performance issue, it is worth referring to two relevant experimental studies by Verlaan et al. (1989) and El-Dessouky et al. (2000). The experiments were performed using air and water at atmospheric pressure and ambient temperature as system fluids. Figure C-4 is taken from Verlaan et al. (1989), and shows the separation efficiency of their studied demister versus gas velocity and droplet size. The mean droplet size was between 5 μm and 20 μm. Figure C-5 is taken from El-Dessouky et al. (2000), and represents their demister performance versus gas velocity and droplet size. Here, the mean droplet size was from 1000 μm to 5000 μm. Comparison between the data of Figures C-4 and C-5 indicates that the provided demister performance data for these two independent studies are in agreement with each other and consistent with the practical experience noted above. In spite of different droplet sizes and apparatus dimensions involved in experiments, the maximum separation efficiency did occur when the air stream velocity was 3 *m/s* in both studies.

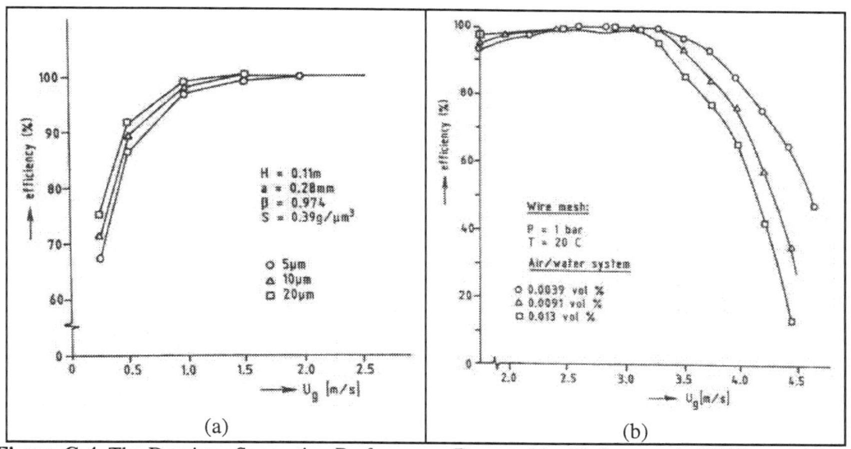

Figure C-4. The Demister Separation Performance Reported by Verlaan et al. (1989) at; (a) Low Gas Velocity, and (b) High Gas Velocity.

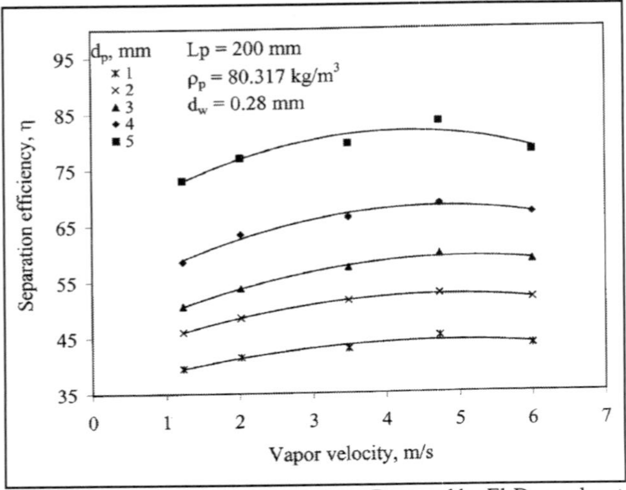

Figure C-5. The Demister Separation Performance Reported by El-Dessouky et al. (2000).

Based on the experimental apparatus of Verlaan et al. (1989), another CFD model including air and water at atmospheric pressure was developed. The experimental apparatus dimensions, with an internal diameter of 39.2 *cm*, were almost the same as the simulated model. So, the same grid system could be used for simulation purposes. The extended water droplet size distribution provided in Table C-1, which resulted in more normal results as per the previous case studies, was again used. The air velocity was gradually increased from 0.02 *m/s* to 4.0 *m/s* and the demister separation efficiency was calculated using Equation C-1. Figure C-6 shows the separation efficiency of demister as a function of gas velocity, which was compared with experimental data of Figure C-4 and Figure C-5. This comparison shows that the maximum demister efficiency and the range of demister operability have been poorly predicted. Table C-3 presents the droplet size distribution in inlet and outlet zones of the demister operating at its best performance, which confirms that demister has successfully eliminated the large droplets from gas stream.

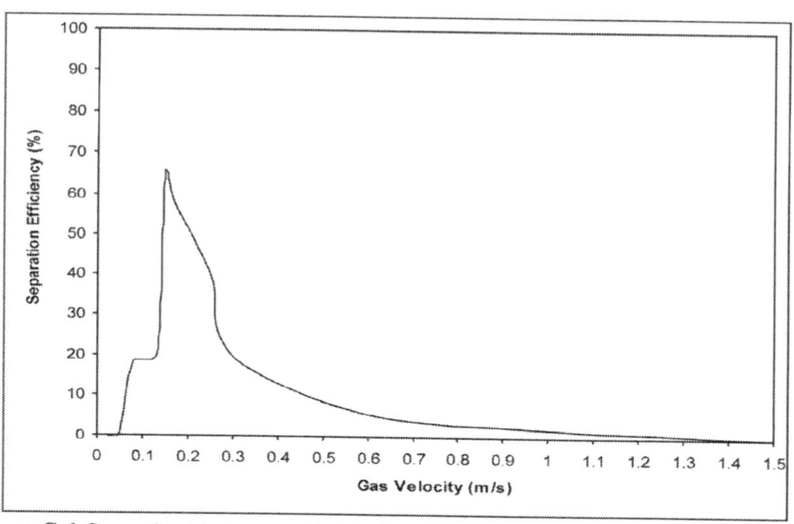

Figure C-6. Separation Efficiency of the Demister versus Gas Velocity for the Air-Water Case Study.

Table C-3. The water droplet size distribution in inlet and outlet zones of the demister.

Discrete Phase Parameters	Demister Inlet	Demister Outlet
$d_{\min}$ (μm)	5	5
$d_{\max}$ (μm)	135	103
$\bar{d}$ (μm)	113	74
n	2.69	2.86

The CFD simulations of mist eliminators are acceptable but there are still shortcomings. As demonstrated in Chapter Four, the predicted pressure and velocity profiles for the demister simulations are as expected. However, the separation performance of demisters has been poorly predicted by the CFD simulations, hence the CFD simulations cannot be used for evaluating the separation efficiency of demisters or their operability ranges. In fact, based on the CFD simulations, the maximum demister efficiency and the range of demister operability are predicted much more conservatively than what is normally experienced. In order to explain these shortcomings, it should be noted that not all the physical properties of demister, as given in Table 3-10, can be specified and modeled via CFD simulations. Therefore, realistic interactions between droplets and demister filaments cannot be simulated properly. Actually, the separation efficiency of knitted wire mesh demisters has been improved by subtle enhancements during the last several years. For instance, as described in Chapter One, in the late 1960's and early 1970's, considerable research led to the design of high-quality demisters through efficient combination of filaments of different materials and diameters (Smith, 1987). The resultant changes in droplet-demister interactions, however, cannot currently be simulated by CFD. Therefore, it is not unexpected that droplet coalescence, for example, with a rate of around 0.1% has been poorly simulated in all demister simulation case studies. Although the overall performance of wire mesh demisters in terms of pressure and velocity profiles and some separation features can be simulated using existing CFD models, the separation efficiency of demisters or their operability ranges cannot be simulated because of complicated separation mechanism involved that are not easily modeled. Consequently, the current CFD models do under-predict the demister efficiency.

Made in the USA
Lexington, KY
01 December 2013